Treasure your memories of growing up.

Gerry Schwarz

PACKIN' CATS FOR THE ARRR-MEEE:

Fun on the Farm in the 'Forties

PACKIN' CATS FOR THE ARRR-MEEE:
───────────────────────────────────────

Fun on the Farm in the 'Forties

by Deanie and Johnnie

also known as

Geraldine Fromm Schwarz and John Robert Fromm

Book design by Jacquie Colvin

PACKIN' CATS FOR THE ARRR-MEEE:
Fun on the Farm in the 'Forties

by Deanie and Johnnie
also known as
Geraldine Fromm Schwarz and John Robert Fromm

published by South Bear Press
2248 South Bear Road
Decorah, Iowa 52101
www.southbearpress.org

First Edition
Copyright Geraldine Schwarz 2011
All rights reserved
ISBN 978-0-9761381-9-8
Library of Congress Control Number: 2011909978

Book design by: Jacquie Colvin
Editor: Dean Schwarz
Editorial Assistance by: Arllys Adelmann, Wilfred F. Bunge and Megan Buckingham
Photographic Assistance by: Eric Erickson, Jerry Grier and Lane Schwarz
Mentor: Roy Behrens

Photographs are from the family archive of vintage and recent photos unless otherwise noted. The tractor illustration is based on a Wikipedia Creative Commons photo at <http://commons.wikimedia.org/wiki/File:Farmall_M.jpg>.

Prepared through Four Colour Imports, Louisville, KY
Printed in China by Everbest Printing Company

Table of Contents

Cats	1	Smokehouse	91
Dogs	12	Playhouse	92
Family	17	Haymaking	96
House	34	Threshing	103
Granary	43	Silo	111
Barn	46	Windmill	116
Horses	56	Trees	120
Cows	60	Snow	126
Steers	63	School	131
Pigs	64	Church	139
Sheep	67	Town	143
Chickens	71	Neighbors	151
Mean	78	War	155
Milkhouse	83	Waste-Not	161
Backhouse	87		

Cats

We were very-very good to our cats—always thinking what we could do to make them more comfortable, to make them happy, to keep them entertained. On cold winter mornings, we liked to have "cat warmings" because we knew the cats would be warmer if they curled up together. We got them all in the woodhouse and made a good spot in the corncobs where they could all sleep. But they didn't stay together very well. The tomcats fought with each other, and most of the others had better things to do. So we tried stuffing them in a cardboard box and folding the lid closed.

Our work always had to be accompanied with a slogan or a song, so pretty soon we were singing, "We're PACK-in' cats for the ARRR-mee." The cats didn't have the same commitment to the war effort as we did—they were not as patriotic. After they had been crammed in the box and escaped a couple times, they really didn't want to stay there no matter how cozy it was. It was kind of hard to catch them again. If they got over the wall between the wood-and-cob part to the coal bin, we gave up on them and settled for any cats we could push in and closed the lid. Then one skinny head would come poking through the little opening and we'd have to start "PACK-in' cats for the ARRR-mee" all over again!

Cat warmings were good Saturday morning projects in the winter, but even better were sparrow hunts in the henhouse. Winter seemed to drive the sparrows into the henhouse where they built sloppy nests from chicken feathers and other dirty stuff above the high windows. They were always flying around in there, crashing into the windows, dropping their dirty manure on us when we fed the chickens or picked eggs. We hated "spatzies," our name for sparrows. So on a Saturday morning we rounded up the best cats, took them in the henhouse, shut the two "people doors" and the little "chicken door" (we forgot to do that sometimes and lost a lot of good sparrows), and we were ready for battle. We also had to stuff boards or cobs in all the crevices around the doors because those sneaky sparrows would find a way out if they could. The cats were usually reluctant at first, but the old-timers—Turk and Big Yellow especially—were excited about the hunt the minute we carried them into the henhouse. As soon as a new cat figured out what the deal was, he was totally into it, too. At first we just pushed the cats up on the rafters, but those were really skinny perches for cats. They fell too often when they leaped after a sparrow, so we poked wider boards in the spaces under the windows and boosted the cats up there. Then we were ready, one of us at one end, one at the other, cobs in hand.

> Throw a cob at the crevices above the windows, scare those sparrows out, keep them on the move. Turk's got one! Get it out of his mouth so he can catch more. Big Yellow has caught one in mid-flight. Ooops! He crashes to the roosts below, but doesn't lose his grip. Good Kitty!

Meanwhile the hens were completely out of control—flapping around when the cobs hit them, squawking when a cat landed next to them, huddling in the back corner under the roost. When the last spatzie was done in, the doors were opened, the hens raced out from their hiding places. The cats and kids were completely satisfied—cats licking their chops, kids replaying each exciting catch.

Mama was not so happy when the egg count was down the next day, but hey, we got rid of a lot of sparrows.

Mice were just as good a treat for cats as the sparrows were. Our really good

Cats were plentiful on the farm, and they made much better playmates than dolls did. But cats have minds of their own and are so difficult to train.

mousing happened only once a year when Lawrence Behm came to shell the ear corn. The corn cribs were full of corn (and mice) all winter. When summer came and it was time to get the corn shelled and delivered and the cribs cleaned up for the next crop, it was a big time for the cats. But the machinery was noisy and it scared the cats. We had to really hold on to them until they saw their first mouse. From then on the cats paid no attention to the noise.

When a crib was first opened, there wasn't much mouse-action because there was plenty of room for the mice to escape the scoop shovels and hide in the corn. But as the corn got down to the last few shovelsful against the sides, especially in just one corner, the mice moved to the walls and started jumping outside from between the wooden slats. In a good crib they poured out of there like water, and the best mousers went crazy. Turk would have one in his mouth and one in each paw. What a cat! Quite a few mice escaped under the cribs, so the cats spent a few days prowling there looking for the ones that got away.

The idea of *buying* cat food never was considered when we were kids. Cat food was mice, sparrows, milk, scraps, and *fish*. One winter when the ice went out of the Shell Rock River in big chunks and threw the carp up on the banks, we launched the "Great Cat Food Project." We went to Chris Carstens' farm with gunnysacks and picked up all the carp we could.

Somehow we made a fish smoker out of a barrel and strung the fish on some Number 9 wire inside the barrel. We took oil that was used for the tank heater and kept the fire going for several days. The combination of fish and fuel oil made a terrible smell in the yard. When the smell got too bad, or the smoking was completed—I don't know which happened first—we took the fish out and hung them by the wire in the box elder trees south of the house. That must have been in March. The carp needed to be kept out of the reach of cats because we wanted to ration those fish for special treats. Either the cats didn't merit enough special treats, or we couldn't get the fish down out of the branches, or we just didn't want to touch them...anyhow summer came with warm southerly

Corncribs were homes to hundreds of mice, and when the corn was shoveled out, those mice poured through the slats and into the waiting claws and teeth of clever mousers.

breezes and an aroma that made a person sick. It wafted right up to the house and through the screen door...an odor that made Mama holler "Peee YUUU! Get that stuff out of those trees!" We thought the fish would eventually fall off the wire and be absorbed into the earth, but that fuel oil smoking must have preserved them. I think those stinking fish hung there all summer.

Another treat for cats was catnip. It grew all over the farm and the cats nibbled it and rolled in it, but all that good catnip out by the woods was just going to waste in the summer, and the cats would really like to have some in the winter, too. We decided to save it for them. We pulled some and hung it up in the granary to dry, but *dry* catnip didn't seem as good as *canned* catnip. So we pulled a big bunch of it, got some empty cans, got good "tamping" sticks and stuffed catnip in the cans—it took a lot of tamping to get a really full can. Well, not too full—there had to be enough room at the top to bend the sides over with the hammer and pound them down tight. In the winter, we said, we would pry those cans open and the cats would have nice fresh catnip. Boy, would they like that! We loaded the little red wagon with our cans and made a pile of them by the corner of the woodhouse, ready for winter. Mama went past there a few days later and yelled, "Peee YUUU! What have you got in those cans?" I think the cans got buried. Too bad, because the cats would have liked winter catnip. And the cans could have been flattened and given to Herman Goldstein, our junk man, for the war effort.

Tin cans were also part of the "Feeding Stations," another in a long line of inventions meant to make cat life contented. For the cats' Christmas we fixed ten tin cans, cut so the cats could put their heads in and drink milk all the way to the bottom, nailed to boards so they wouldn't tip over, positioned in front of the cow stanchions for easy feeding at milking time. They were supposed to help the cats stay cleaner. That old galvanized pig pan we had been using was too wide and some cats stood in the middle of it when we poured the milk in. They were so anxious about getting their share that they fought with each other, and some always stood right where we were pouring, got milk in their faces, on their backs, and of course all over their feet. Not good to have wet fur and feet in the winter!

The new and improved "Cat Stations" would be so much better. A place for each cat, no fighting, no milk on the backs, no feet in the milk....

But getting the milk into the cans was a little tricky. Quite a lot went on cats' heads as they dived in for the first drops. Some went between the cans and froze there. Frozen milk in the winter wasn't so bad, though. The cats could lick it between feedings, which gave them something to do. Then summer came. And flies. We took the contraption out by the tank and threw a pail of water on it, but it was not really sanitary. It never really smelled sanitary either. But it was our invention and the cats had to learn to live with it.

Mama always said, "No cats in the house!" So of course we tried to bring them in—using stealth and disguise. "Just a box of old rags," we said as we walked through the kitchen with a box wobbling in our arms, one hand holding the "rags" down. "What's that tail doing hanging out of your box?" So we had to take the "rags" back outside. Sometimes we were lucky and got all the way into our bedroom. We put the box under my bed and I held the cat in there until Mama had said good night and left the room. Then I pulled the cat up through the headboard slats and under my covers. About that time Mama would come back in our room. "I think there's a cat in here," she'd say.

"No, no, it's just us." We made loud purring sounds to cover up the cat's enjoyment at being under the warm blankets. If Mama left the room soon enough, the cat wouldn't get worried about that close confinement and start yowling. Otherwise we had to change our purring imitations to yowls and meows. Eventually we had to give up our cat, hand it over to her, and go to sleep.

We thought of another way to get the cats in the house. I'd go upstairs to my room, open the window, lower a clothesline rope down to the sidewalk while John would tie the rope onto the handle of a grape basket, catch a cat, put it in the basket, cover it with a flour sack tucked in all around, and signal me to pull the cat up. I don't think we ever got one all the way up to the second floor. If we got one to the top of the back door, we were doing pretty good. Good for us, not so good for the cat. No matter how fast I pulled, the cat worked its way out from under the flour sack and leaped out—an eight-foot drop at least. After a couple falls like that, a cat was really wary about being caught again. I suppose we could have opened a downstairs window and let a cat into the house that way, but where would be the adventure in that?

Cats were a great education for us. Cats taught us to be inventive, persistent, thoughtful and (most of all) kind. Even when we bent their ears backward and said, "Look, Kitty has no ears...Kitty has one ear...Kitty has two ears," we were not being anything but kind. The cats seemed to like it. As for removing loose whiskers—if a cat has a loose whisker he won't be worth anything all day—that was a real service to the cat.

Sunday dinners often meant Aunt Mae and Aunt Clara sitting around the table,

8 PACKIN' CATS FOR THE ARRR-MEEE

Skippy was a bottle-fed kitten that developed rickets and walked with a curious gait. Rusty was a great collector of compliments. They both craved human affection.

---

sucking their teeth and talking about our ancestors and all their descendants. Genealogy seemed to us to be an important topic—for our cats. We started a half dozen different Cat History records using those farm record books sent out by lumberyards or machinery dealers.

Each history started out with the first two cats we could remember: Mitzy and Mussy—the great-grandparents of all future cats. Each cat family had a new page, and every kitten in each litter was named, sometimes posthumously. We knew that Mitzy had five kittens in that first litter, but only one lived long enough to get a name—Annabell—so we named the others a few years later.

We gave our cats creative names, too. Little Cat was one and My Little Cat was another. We had lots of Yellows—Big Yellow, Little Yellow, Long Yellow, Wild Yellow, Yellow Yowler. The other colors were also names—Blackie, Whitey, Grayie, Buffy.

Trousers was named because we were singing "Bell Bottom Trousers" when we found him and his litter mates in a nest in the barn. Skippy was named after peanut butter, but it was a good name for him because he'd been bottle fed after his mother died. He developed rickets, so he walked in a really strange way. Thunder and Lightning were carried to one of the basement window-wells by their mother during a rainstorm. They didn't last long.

Mange (what an awful name) lived a long and useful life as a mouser and mother, even though nobody liked her much. But most of the cats didn't get through a full year. In the "Cause of Death" column in the Cat Record Book, we listed "Plague" for most of them.

```
CAT RECORD CO.
           RECORD OF 1948 -1947
    All most all of-cats-the cats are dead or missing.
    Wimpy;s family is dead, all but 2.
    Grayes family is alled dead.
    So is MITZEES family all dead.
    A bad year if you ask me.
```

*John
Very good why hide your talent
write up some more things like it!
L. F.*

One Christmas we got an almost-toy typewriter and spent many hours typing the Cat Newspaper. At last this article by John will be published.

Once when Doc Spearing came to vaccinate pigs, we asked him why so many of our cats died of the plague. "Oh, that's distemper," he said. "If you can catch all of them and put them in a crate, I'll vaccinate them, no charge." We asked why they had to be put in a crate. "Well, so we can be sure they're all here and each one gets vaccinated." We were offended. Didn't he know that we didn't have to catch our cats? Every cat on the place came right to us when we gave the All-Cat-Call: "Kit-EEEE, kitty, kitty, kit-EEEE!" And we certainly could remember the names of which cats had been vaccinated—there were only about twenty of them.

The cat record book led to our gravestone project. When the new house was being built, some cement was usually left over after the men finished work at the end of the day. We built little square forms, filled them with cement and wrote the cats' names in the wet cement. The names were of the living cats because we weren't sure where all the others were buried, even though the cemetery drawing in the back of the Cat Record Book showed several graves, including one for our dog Rex. Two of our cats were named Amos and Andy, one of our radio programs. Maybe we didn't think about the hired man's name when we named the kitten. After we made that gravestone, we noticed Andy, the hired man, walking around it a couple times, looking at it. He may have thought his days were numbered and he was headed for the cat cemetery next to the playhouse.

RECEIPTS AND EXPENSES

The daily account to be kept on pages 6 to 12 should include all cash transactions of the year that belong to the farm business. All such items as interest on notes and mortgages should go in as regular expenses. At the end of the year all expense and receipt items should be classified and entered on the summary pages 26 and 27. An example of several entries has been inserted in the first few lines of this page. Keep accurate detail of expenses for feed, seed, fertilizer, etc., in order to determine profit or loss on individual farm operations.

DATE	ITEM	QUANTITY	PRICE	AMOUNT RECEIVED	AMOUNT PAID
Jan 7	Repairing Wagon – Christ Meyer				5.00
" 11	Liberty Elevator Co.	100 Bu. Wheat	1.10	110.00	
" 18	Lumber for Shed				31.50
" 29	Blair Hdw. Co. for Nails				7.80
Date	name		born	died	no	died of
	Jim & Jo's Family					
July 4	Buffy		1946		1	
"	Whitty		1946	1948	2	Gun
"	Mama's cat		1946	1947	3	Plague
Apr. 6	Big Yellow		1947	1948	4	"
"	Owl		"		5	
"	Calico		"	1947	6	"
"	Freckels		"	1947	7	"
Aug 17	Skookom		1947	1948	8	"
"	Pet		"	1948	9	Tank
"	Big Tank		"	1948	10	Plague
"	Big Jewel		"	1948	11	"
"	Splendid Emper		"	1948	12	"
"	Light Yellow		"	1948	13	"
"	Twinkle		"		14	
Apr. 26	Skippy		1948		15	
"	Toots		"	1948	16	"
"	Tippy		"	1948	17	"
"	Ricky		"	1948	18	"
"	Lula		"	1948	19	"
"	Mewey		"	1948	20	"
NR	Wimpy		1944	1948	21	"

We tried to keep up with cat genealogy as best we could, but many entries simply said NR (no record). Not all cat deaths were from the plague, but who would have shot poor Whitty?

FUN ON THE FARM IN THE 'FORTIES

Dogs

We always had one major farm dog. The first one I remember was named Major, kind of an old-fashioned shepherd, not one of those fancy police-dog German shepherds. We petted him when we gave him table scraps, but he wasn't a pet. He was a working dog. He brought in the cows at milking time and brought in the steers when they had been out in the field cleaning up the stalks after corn picking was done. Pa would climb up on a gate, wave his arm and holler, "Way 'round 'em. Way 'round 'em." Major ran back and forth behind the cattle, barking and nipping and bringing them closer to the barn, for as long as he could hear Pa hollering and see him waving his arm.

When Major was old and gray around the nose, we got Rex (pretty much the same kind of dog), so the old dog could teach the new dog. Rex was a good cattle dog, too, but a boring dog for kids to play with. By the time he finished a workday, he was just about played out. If Pa was doing fieldwork, Rex ran along near the front of the team or the tractor, row after row, sniffing and wagging his tail and chasing a rabbit now and then.

Duke, Wolf, Jip and Manfred each served a term as farm dog. But we had quite a few auxiliary dogs, too. When the neighbor dog had pups, John and I each got one. John's was black—"Blackie." Mine was brown—"Brownie." We were very creative about names. We played with them under the walnut tree, gave them rides on the swing—sometimes on our laps, sometimes alone on the

Puppies were great to play with in the summer, but as fall came and they got older, they seemed to disappear. Maybe that look on Deanie's face hints of things to come

swing board. They weren't crazy about that. We gave them rides in the coaster wagon, too. They weren't crazy about that, either. We taught them to swim—in the cattle tank. We took them down to the barn every day for their swimming lessons and tossed them in with the green algae and floating slobber from the cows' mouths. My brown pup went in, reluctantly, every day. John was kinder to his pup. He called it "the dry black one."

Pa and John and Heiny. All three chased chickens from time to time but only one was banished for it.

After summer was over and we went back to school, we didn't have much time for the pups, so it may have been four or five days before we missed them.

"Hey, where are those pups?"

"They went for a ride with Short."

Short Heidenreich was our hired man and later our neighbor with his own farm down the road. We imagined our pups happily riding somewhere nice with him.

Heiny was an interesting "extra dog" for us to play with. He was kind of homely, but he was lively and more fun than Rex because he ran around the yard with us and chased us. Unfortunately he chased chickens, too. And caught them. And ate them.

"Hey, where's Heiny?"

"He went for a ride with Short."

We could have had a lot more puppies if Uncle Delbert had really meant that he would bring us a puppy. He and Aunt Aurora (we thought her name was "Roar") lived in Minnesota. Every time he came to the farm, he said, "You kids didn't want a puppy did you? I saw the nicest little white puppy by the side of the road, lost, and I thought about bringing it to you, but I didn't think you'd want a puppy." We'd stand there with such sad faces. Of course we wanted a puppy! "Well, next time I come...." Next time he would say, "I was at this farm and they had a whole lot of nice little speagles. I was going to bring you one, but I didn't think you'd want a speagle." Sure, we wanted a speagle. "Well, next time...."

Rusty was a city dog—a cocker spaniel—given to us by someone who was moving away. We felt pretty smug about having a dog with a real breed name, and he came complete with a real collar. We took him to Mason City with us and took turns walking down the sidewalk with him on a leash (no, not a real leash—a short piece of clothesline) and collecting compliments. "Oh, what a nice dog." "Pretty dog." "What's his name?" all counted as compliments. I made the mistake of walking into the Piggly Wiggly grocery store with him and found out that grocers don't give compliments to dogs.

When Annie Pryor died—she who kept a flock of chickens in her basement—we inherited her dog Trixie. For one day. He ran away, probably trying to get back to Annie and the chickens. We felt terrible. And even worse when Short told us he found the carcass in the fence over east, the collar caught on the wires. Oh, poor Trixie.

Now Wolf! There was a dog! A white German shepherd—friendly to us, nasty to strangers ("Don't get out of the car 'til I get aholt of the dog...") and death to rats! When we ground ear corn for the cattle, a little corn always spilled off the scoop shovel as it went from the back of the wagon into the grinder. There was a metal sheet, maybe four feet square, on the ground between the wagon and the mill, which made clean-up easier. But some corn rolled off the edges of the metal. Rats tunneled under it and carried the corn back to their nests, so they were lying-in-wait while the mill was running.

But Wolf wasn't lying-in-wait. He was prancing around, anxious for the fun to begin, or standing tensely with his nose to the edge of the metal sheet. John egged him on, "Ready, Wolf? Are you ready?" In a flash John lifted the metal sheet and Wolf was on those rats, grabbing and killing one after another as they ran for dear life in all directions! What fun for all of us (except the rats).

Family

Johnnie and I (Deanie) were both born during the Great Depression when farming was chance-taking and hanging onto our farm was "no sure thing." We began to learn about life beyond the farm during the early 'Forties when winning the War was "no sure thing" either. These worries swirled around in the grown-up world above us, but down in our kid-world, life was good and predictable and full of things to do.

When Mama married Pa (she was 22, he was 23), she moved from her folks' home in the thriving town of Plymouth to the Lime Creek Township farm three miles away. The twelve-room farmhouse was full of Pa's family. Gramma Fromm was the head of the household then. She gave the orders and kept everybody at work. She had inherited the farm when Grampa Fromm died. He was a German immigrant who had fought in the Civil War. When he was a teenager on his family's Wisconsin farm, he received one thousand dollars (some say five thousand) from a doctor to take the place of the doctor's son in the army. After the war, and after he had recovered some of his strength from dysentery, he moved to Iowa and bought the farm with that money. He coaxed his brothers to move, too, and three of them bought farms just up the road. Two other brothers couldn't come because their wives were afraid to cross the Mississippi River. At Sunday dinners, our old aunts would tell about going to visit their Wisconsin relatives. They all sounded very rich because they had huge weddings and funerals. We heard the story of Charlie Neiman's extravagant Golden Wedding Anniversary

18 PACKIN' CATS FOR THE ARRR-MEEE

Charles Henry Fromm—ready to motor to Plymouth to take flowers to his girl, Gertie Holcomb.

over and over. Some of the cousins had raised ginseng and silver foxes. Wow! We tried to imagine a fox that was silver.

When Grampa Fromm was nearly 30, he married Ann Kinney, a local girl who was 16. Sometimes he would declare to his wife, "I was marching across Georgia when you were just learning to walk!" Their seven children all spoke English because Gramma insisted—her ancestors were English. The other brothers and their wives spoke German. Their kids spoke "broken English," which Gramma made fun of. Those wives were homesick for Wisconsin and their families, so by the time John and I came along, they had all moved back. Pa had grown up with lots of aunts, uncles, and cousins, five older sisters and one younger brother.

We liked to think about our Grampa Fromm, how he must have felt so far from his folks, buying his own farm when he was only twenty-five. He worked so hard plowing land that had never been plowed and planting trees to make it feel more like Wisconsin or Germany. How glad he was to find such nice level land and live in a sweet country like Iowa. He must have been proud when he looked at all his fields and new buildings and cattle and pigs and chickens. Mama told us how much he liked the wild phlox that grew in the ditches, so every year on Memorial Day, we picked a big bouquet of them and put them on his gravestone. Grampa Fromm died when Pa was 18. This meant that Pa didn't get to finish high school before he became the farmer. With Gramma giving orders.

When Mama came to the farm, Gramma Fromm was in charge. Mama made up her mind that she was going to get along with Gramma. She really did get along better than any of Gramma's five daughters did. Gramma told her she often didn't get off the farm for nine months straight. Mama could drive a car, so she took Gramma wherever she wanted to go—to town, to visit her relatives and even to Clear Lake to put her feet in the water. Gramma's bachelor brother, Uncle Charlie Kinney, lived in the house, too, along with Pa's unmarried sister and brother, a hired man or two, and usually a married sister and her family who came for a nice long summer visit. The first time Mama and Pa were alone in the house was the night our brother Charles was born. That was when they really needed some of those relatives.

By the 1940s the older generation had passed away and our family life centered around our folks. Pa (Charlie Fromm) was always kind and good to us kids, calling us little pet names—John was "Jackson" and I was "Jeannie with the light brown hair." (It was actually red.) When rationing ended after the war, Pa could buy gas and tires and even a better car. He took us along to things he thought would be fun for us, like farm auctions, the salebarn, band concerts in the band shell at East Park and especially (on Sunday afternoons) Bayside, the amusement park at Clear Lake. There he watched us bash into each other with the Bumper Cars. Going to Clear Lake was a huge treat for John and me. We vied with each other to be the first to see the lake, and when Pa drove past the slough on one side of the lake, we always asked each other, "How would you like to live in there with food and water?" The Outing Club was a fancy place with a long row of "'tached-on houses." We *really* wondered about living in there with food and water.

Some evenings Pa took us to the Manly depot where we watched the Rocket come sliding smoothly up to the station. We liked to watch people getting on and off with their suitcases and see the folks dining in the dining car. We thought … maybe someday … He drove us over to the railroad tracks a mile east of the farm to watch the Rocket go by at eight o'clock on a summer evening. Sometimes we saw a freight train with a rumbling black steam engine that spewed wonderful-smelling black smoke and shrieked a whistle that rattled the windows in our house. Those engines pulled cattle cars and box cars and cars full of coal and grain. We counted each car slowly, together, and looked carefully into each open boxcar, hoping to see a bum catching a ride.

In 1939 Mama is holding the squirming two-year-old Deanie. Marilyn (15) and Charles (19) stand at attention. Pa, tanned from long summer days in the field, holds a deceptively innocent-looking Johnnie.

Pa worked hard every day. Farming was a lot of handwork and legwork, with feed to shovel, manure to pitch, water to carry, horses to hitch, cows to milk. He was tired by the time supper was over and we were ready to go to bed, but he was not too tired to read to us every night—maybe a chapter from one of the Thornton W. Burgess books like *Old Man Coyote* or *Happy Jack the Squirrel*. Some of the stories were scary and made us hide under our covers. When Pa read about the little animals finding a half-eaten chicken left by a fierce stranger in their forest, we were terrified. And when that fierce stranger—Old Man Coyote—chased Peter Rabbit, who ran lipperty-lipperty-lip to the dear old briar patch, shivers went up our spines. It was fun to be scared, especially when we were snug in our beds, and Pa was nearby. We begged him to read that chapter again, night after night.

Pa usually went to church on Sunday evenings, and then Mama would read to us. But it was hard to go to sleep until we saw his car lights reflected on the wall of our bedroom. Then we knew Pa was home, and we felt safe.

Mama before she was Mama—on the south side of the Plymouth school where she was teaching eighth grade in 1918.

He liked to tease one neighbor who wasn't much of an early riser. Pa sometimes called him first thing in the morning: "I just came in from getting all the morning chores done. I suppose you've been out and got the milking done by now … Oh? … Oh? … Well, I hope I didn't wake you up or anything."

No one at our place ever used swear words. Not even the hired men and the neighbors used swear words within our earshot. If John or I said Gee or Gosh, we were quickly corrected, because those were "bywords"—substitutes for

Jesus and God. We didn't use four-letter-words either—we used farm-words. When John was about two, Pa and the preacher were visiting under the walnut tree in the yard. John was saying, "Pig manure, cow manure, chicken manure, horse manure…." The preacher asked, "What's he saying?" Pa said, "Nothing. He can't talk yet."

And of course Mama, Gertie Holcomb Fromm. She ran everything in and around the house, including us. Mama had been a teacher in country schools and at the Plymouth school until she married, so she was always teaching us. In fact she home-schooled Charles in first grade before "home-school" was even a phrase. She got the readers, spellers and arithmetic books and taught him, because it was fun and because with his June birthday, he was pretty young to start school that September. Marilyn had an August birthday, but when she was barely four, Mama decided that she was old enough to start school.

When Mama told Pa, who was then nearly 41, that a new baby (me) would be born about the middle of January, he just put on his cap and went outside without saying anything. He was working hard to provide for three kids during the Depression; the thought of one more must have been overwhelming. But on that January day in 1937 when he stopped at the country school to give the news to our sister Marilyn, he was all smiles. "We have a long-legged, red-headed baby girl!"

Mama was particular about grammar and pronunciation. "Don't end your sentence with 'at.' Don't say 'should have went.' Don't say 'ain't.'" When John began saying, "I seen" (maybe just to be obstinate), she clamped down on him. She taught us not to pronounce the "t" in *often* but to be sure to make the "th" sound in *three*. (She shook her head when she remembered her students at a country school in South Dakota who couldn't say *three*.)

Mama was also particular about our handwriting—she worked on her own penmanship and compared it to letters her father had written, which were

> Artie Has a Tile factory runing 3 Presses with a Man at each Press Making 1200 Tile Daly His office is 14x16 Richly furnished He also Has 4 men Laying Tile Besides a Sailsman & Colector His Payrole is Between 6 + 700 d week

Mama admired her father's handwriting.

> B.M.   Feb. Thur. 16th '78
> Yes, I said to Myself you rem'd the Feb 8 birthday and so I started looking thru chk's (being in the I.T. tax making just recently) & sure enuf a Chk there for Feb 8 — May 9 — But 1977 instead of 1978 — Seems sort

She was careful about hers and critical of our poor attempts.

beautiful. She taught us the names of the weeds in the garden and the flowers in the woods. And she taught us the words to the songs the birds sang—the meadowlark sings "I want to eat with you," the robin sings "Cheer up, cheer up." The bluejay sings about himself, "Thief, thief," and the goldfinch sings about himself, "Sweet, swee-eet." The bobolink and chickadee both sing their names, and the barn swallow sings, "Quick, quick, hurry, hurry." The crow doesn't sing—he just says "Caw, caw, caw." Lots of people didn't like crows, but Mama did, so we kids did, too. Mama even named one kind of birds—the plowbirds. Flocks of big white birds came in the fall to our freshly plowed fields, covering the black dirt like snow and feasting on the bugs and worms that the plow turned up. Later we found out they already had names—terns—but we still always called them plowbirds.

Mama kept bars of Cashmere Bouquet hand soap in her dresser drawers, so her clothes carried that scent in the morning, before the farm smells and kitchen smells took over. We noticed that sweet smell especially at church when we leaned against her and drew tiny pictures on the edges of our Sunday School paper. For that hour we were good little children and sweet, too.

Even though Mama was a "town girl," she had always milked cows and raised chickens in Plymouth and peddled milk and eggs to folks there. When the Great Depression came and we almost lost the farm, she did her bit to add to the farm income by going door-to-door in Mason City and setting up a peddling route. Then she baked apple pies and picked fresh vegetables from the garden and sold eggs. She raised extra chickens, butchered and dressed them and peddled them, too. When better times returned, she kept on baking and butchering because she still liked peddling. One day she overheard a customer telling a neighbor, "I don't think Mrs. Fromm really needs to be peddling anymore." So she quit.

But Mama was always working. Summer and winter, she was up before daylight. In the summer she was in the garden early in the morning, planting or weeding

Mama outside her kitchen window. She always wore an apron over her clean, pretty cotton dress, but she took her apron off before she had her picture taken.

Right: Marilyn, Johnnie and Deanie—all washed up and ready for Sunday School.

or bringing in the vegetables. We each had our own little Victory gardens, but we weren't very industrious about keeping them clean. She was. In fact everything Mama was in charge of was clean. Dirt stayed outside. The carrots and beets and potatoes were washed under the pump by the back door. They didn't come into the kitchen until they were ready to be peeled or boiled or canned. Canning the garden produce seemed to go on all summer and long into the fall. John and I helped some, but since it involved boiling water and very sterile jars, our hands were not especially suited.

Monday was washday, and that was a major event. After breakfast we brought the two rinsing tubs in from the porch and helped fill them with water. In the early 'Forties, before water was piped into the house and a water heater

installed in the bathroom, that meant pumping and heating water for the Maytag washing machine and the rinsing tubs. Mama made her own soap from beef tallow and lye. Since it was made only once a year and cut into squares that became hard and dry, she had to put it in a pot to soak overnight or set it on the cookstove first thing in the morning so it would be soft enough to make the washwater soapy. When the clothes were hung on the line, they smelled clean, but not sweet.

Tuesday was ironing day. I tried to help Mama sprinkle the clothes and roll them up so they would iron more easily. It was hard to get just the right amount of water in my hand and shake it evenly over the pillow cases or Sunday shirts or my dresses. Mama bought a sprinkling plug that fit in a pop bottle. Even that was tricky for me. More than once I shook it so hard that the plug flew out and the shirt was drenched. Ironing was way beyond me until Mama got an electric iron after the war. Even then I seldom ironed a dishtowel without scorching it.

On Wednesday, Thursday and Friday, Mama had any number of work projects: sewing, baking, cleaning. Sometimes it was fun to help Mama clean. She had a carpet sweeper—a long-handled brush that rolled over the floor and flipped the dirt into a container on top of the brush. We raced back and forth with it over the carpet in the parlor. We liked to use the dust mop, too. A rag with a little oil on it was fastened to a long handle. We swished that back and forth under the dining room table and under the rocking chairs. I didn't want to swish it under the beds, though, because there were boogle-degaws under there. Years later, I heard someone call them "dust bunnies," and that sounded much less terrifying. John and I liked to be in the house when baking was going on. We licked out the bowl after the cake batter was poured into the pans. We fought over it until John reminded me that there were raw eggs in that batter. Eeuw. He could have the rest of it.

Saturday night was bath night. Of course, we took baths more often than that because Mama told us that her students in South Dakota smelled so bad by

Summer meant lots of company on the farm. Some came for several days; some came for several weeks. In this 1937 photo, Mama is holding Johnnie, Cousin Henry Haegg is next to Charles, Aunt Bertha Haegg is holding Deanie, Aunt Mae McClintock is behind Cousin Norma Lou Haegg who is peeking around Gramma Holcomb. Marilyn is pressed against the playhouse. Behind the group is the woodhouse, which would become the scene of many cat warmings in future years.

Friday that she could hardly get close enough to their desks to help them with arithmetic problems. But on Saturday nights Mama washed my hair and sometimes set it in rags so I would have long curls for Sunday School. Other times she braided it, hoping it would stay in place overnight. She combed and parted John's hair, but it was always mussed up right away. We were so lucky to have a bathtub and hot water. Marilyn told us how it was when she was little before the second set of stairs had been taken out and a bathroom put in that space. Baths back then were right there in the kitchen in front of the cookstove where the water was heated. And the house was full of people always tromping through the kitchen.

By the time John and I were able to remember much, most of those old relatives had moved away or died. The people at home were Pa and Mama, we four

Charles Walter Fromm with his 4H Angus calf—together they won the purple ribbon at the North Iowa Fair in 1938.

kids and usually a hired man. The oldest kid was Charles, 16 when I was born. He liked to tease me about things I was afraid of, like people with "cutted off" hands or fingers. He'd bend one of his fingers under and say, "Look, Deanie, I've got a cutted off finger." "Waanh, waanh," I'd run crying to Mama and bury my face in her apron while he laughed and laughed, and Mama scolded him a little. Other than that, he never did anything bad. He was a smart student, a star in 4H—taking the purple ribbon at the North Iowa Fair with his Angus calf—what a high point that was for our family! Years later I heard friends of our family say that for our folks, the sun rose and set on Charles.

Even though the subtitle of this book is *Fun on the Farm in the 'Forties*, 1940 had a sad beginning. That January our whole family was sick with the flu and Charles died right there in his bedroom next to the room where we were sleeping. The next morning when I went into his room, his bed was made and empty.

I asked, "Where's Charles?" Mama started crying. That was the first time I ever saw her cry, and it scared me. I didn't ask any more questions, and I don't think anyone actually told me what happened. Mae Reindl came to our house and stayed with John and me while the others went somewhere—later I figured out it was to the funeral. Mae was crying, too, so there was no use asking her what was going on. Mama didn't go to church with us for a long time after that. All our places at the supper table were shifted around a little so that his empty place didn't exist. Marilyn moved from her room upstairs and slept in his room. No one ever talked about Charles again, at least not when Pa was around. All the pictures of him were put away in Mama's cedar chest. I didn't understand how much pain the loss of his firstborn caused Pa. Charles was good at everything. He'd become an Eagle Scout at sixteen; at nineteen he was in his second year at Junior College. He could have gone anywhere, done anything. His future was so bright. We didn't remember clearly the years before that, but when we were grown up, Marilyn told us that Pa was never again the light-hearted and cheerful father she had known.

Marilyn was fifteen at that time, and she made it her duty to have suppertime be pleasant for Pa and Mama by telling interesting and funny things that had happened that day. At school she tried to find things she could talk about at supper. She read to us at the table and cut funny cartoons out of magazines. Maybe it helped. We didn't know what she was doing; we just thought she talked a lot.

She was also a model child, someone held up for us to be like—she did everything well. Mama told me she always put her dolls in their beds every night and covered them with blankets—mine were scattered around the room, naked. Her room was always neat; her grades were perfect. She was always kind and good to us, no matter how mean we were to her. She entertained us by telling long and involved made-up stories about people we thought we almost knew. She tried to make our meals exciting by pretending to be a waitress, putting the food on our plates "just so," and calling it "the blue plate special." She showed us how to take better care of our fingernails, to file them instead of biting them,

Pa as we remember him best. His "overhalls" have been mended many, many times, and the straps seem mismatched, but he would be ready for a long day of work as soon as he put on his cap.

---

to push the cuticles back (ouch) and to clean under them with that pointed file, "like this." Yikes, that hurt! She wanted to get every last bit of dirt out, and some of it had probably been under there for days.

She could talk in public, so she was an outstanding member of any group she belonged to—Sunday School, Methodist Youth Fellowship, 4H. (I flunked 4H.) She did such a fine job of demonstrating the care of nylon hose as her 4H project at the fair that she was hired to work at Merkels Department Store! We were so proud when we went into Merkels, and there was our sister putting the customer's money in that little capsule that whished up a tube to the office and then, by the time she had wrapped up the customer's package, whished down again with the receipt and change. Fascinating.

After Junior College, Marilyn went to the University of Iowa. One of her teachers was Marcus Bach. His class lectures were broadcast on WSUI, which Mama could just barely get on our kitchen radio. One day he introduced his student, Marilyn Fromm, and she read one of her papers on the air! Pa said that Mama's feet never touched ground the rest of that day.

House

The best way to go into the house was through the back porch and into the kitchen. The best room in the house was the kitchen because Mama was most always there. The kitchen had big windows on three sides, so it was usually sunny. It had a big cookstove so it was usually warm. In the southeast corner was the kitchen sink with a cast iron hand pump for soft water from the cistern and (after 1945) a faucet for hard water from the well. The faucet was high enough above the sink that we could bend our heads and drink directly from the stream of water—no drinking glass needed. We had hot running water if Mama ran fast enough from the cookstove to the sink with the teakettle to rinse the dishes (an old joke that got pretty tiresome after a few tellings). John and I did dishes and argued about which was harder, washing glasses or washing silverware, and which was better, washing or drying. I usually had to dry. John finished washing and dashed outside while I still had all the silverware to dry.

We ate every meal at the big round table in the kitchen unless we had threshers or aunts and uncles for a meal—then we ate in the dining room. We felt pretty fancy when Mama got a new oilcloth from the dime store for the kitchen table, once a year or so. It had a strange smell that wouldn't wear off for several weeks, but we still liked it because it had new patterns, new flowers, new colors.

This is the farmhouse where Pa and his sisters and brother and his first two children were born. Johnnie and Deanie spent the 'Forties living in its twelve rooms. In 1905 the professional photographer posed Gramma Fromm, Aunt Mae, young Charlie Fromm (Pa) with the dog, Uncle Charlie Kinney (Gramma's brother), and Max Hacker (who may have been Aunt Mae's "fellow").

At one time there was a stairway to the cellar and to the upstairs on the south side of the kitchen. When the house was "modernized," the steps to the cellar were covered over and a stool and bathtub put in that space. The door that once led to the upstairs had the roller towel on it. We washed our hands in the sink and dripped across the floor to the towel, hoping to find a spot that was clean and dry. The steps going upstairs had been taken out, but just inside the door were several shelves for baking supplies. I really liked to eat some of the baking chocolate or cinnamon sticks. If no one was in the kitchen, I would get a toehold on the lower shelves and climb up to the shelf where they were kept. Once, just as I was reaching for the chocolate, Pa came in from outside. "Don't tell Mama! Don't tell Mama!" I bawled. Mama would have spanked me, but Pa never gave spankings. I lived in fear for the rest of the day, but he must not have told on me.

The telephone was fastened high on the kitchen's west wall. I had to stand on a chair to talk into the mouthpiece, but I seldom talked on the phone—kids didn't have anything that important to say. We were on a party line, of course, with 12 other families. Even though the phone rang several times a day, it was not often with the three-longs-and-a-short that meant someone was calling us. When we heard a different ring, sometimes we "listened," but if we weren't really careful lifting the receiver, the people who were talking might say, "I think someone is listening. We better not talk anymore about you-know-what."

Part of the wall between the kitchen and dining room was a cupboard that had drawers for silverware, tablecloths and napkins that could be pulled open from either room. Above them were shelves for the plates and cups and other dishes. There were cupboard doors on each side—tongue-in-groove doors in the kitchen and glass doors in the dining room. Mama put the pretty green glass plates and other fancy dishes on the dining room side.

The dining room was a fine large room with a couch next to those cupboards. Pa took his nap there after dinner at noon, and he usually had a toothpick in his mouth then. When he finished using it and laid down on the couch, he pushed the silverware drawer back a little and dropped the toothpick in the space underneath. Once John pushed that drawer back and found hundreds of chewed toothpicks.

Next to the head of the couch was Pa's big radio, which later was adapted to have a record player run through its speakers. Pa liked "good music"—Beethoven, Brahms, Mozart…and John Philip Sousa. He listened to his records on winter Sunday afternoons. (John and I didn't care much for music that didn't have any words to it.) Two rocking chairs were also near the radio. In the evenings Pa sat in one and listened to war news and read the newspaper. Sometimes Jim Otzen came over in the evening, and they sat there and told stories. From our beds in the next room, John and I hollered, "Talk louder! Talk louder!" We didn't want to miss out on Jim's stories.

This same roller-and-towel hung in the kitchen several steps away from the sink. By the time we had drip-dripped across the room not much water was left for the linen towel to soak up. Because we could use so many different places on the towel, it didn't need to be changed every day.

Of course the dining room had a table, square with big heavy legs. When threshers were coming, we pulled it out and put in all the leaves, so that at least a dozen hungry men could sit there. Then there were even more stories! Two other wonderful inventions were in the dining room: a wind-up Victrola and a heat register on the floor just above the furnace. We learned to wind up the Victrola and set the needle on the records when we were very young. We knew the words to all the songs in the old brown albums of records, and we sang all the time when we went around the farm. "You are my Sunshine," "Way down upon the Swanee River," "Old Dog Tray"....

The register on the floor was our favorite place to get dressed in the morning. John and I jostled for the best position and did some pushing and shoving. A children's morning show that we liked to listen to on Pa's big radio featured a getting-dressed race between the boys and the girls. The announcer gave a play-by-play description of who was winning. "The girls look like they are a little lazy this morning. Suzie can't figure out which is the front of her dress and she has it on backwards. The boys are almost finished, but Johnnie has buttoned up his shirt wrong. Start at the bottom, Johnnie, and you'll come out even at the top." That little man inside Pa's radio could see everything we did. He was right! Johnnie HAD buttoned his shirt wrong.

The heat from the register could be controlled with a chain loop fastened to the wall. If we pulled it one way, it opened the furnace damper and we got more heat. Pulled the other way, it closed the damper and the heat stopped. Sometimes, if we had company that stayed too late, Pa shut the damper and the room cooled down. When the company got cold enough, they put on their coats, and Pa would stand up and say, "Well, it looks like you're getting ready to go home."

The other rooms downstairs were in the old part of the house—well, the whole house was old, but the kitchen and dining room were a little newer. Three doors were in a row on the north side of the dining room. The left one opened to two bedrooms. We had to go through the first bedroom to get to the second. Four of us slept in the first one, Pa and Mama in the big bed, John in his youth bed and I in my metal crib. Marilyn usually slept in the other bedroom—she was ten years older than John and needed her own space.

The middle door opened to the steps going upstairs, a place with infinite possibilities. The third door opened to the parlor. A Round Oak wood/coalburning stove was in that room, but the parlor was hardly ever used. Once in a while old relatives from Minnesota or Wisconsin came to visit, so they sat in the parlor to talk and have coffee. I was scared of old relatives, so I hid. That alone made the room scary, but there were also big paintings done by Aunt Kate that hung on the walls—one was of a lion and a dead cub titled, "The Death of the Firstborn." So sad and so scary. Another one was of a waterfall and another was of a creek running over a log. I liked those better. Still, I'd just as soon stay out of the parlor altogether.

Upstairs were seven bedrooms, all with double beds that sloped toward the center, ("deep valley beds"), dressers with drawers that stuck and clothes that no one had worn for years. Each room had stories about whose room it had once been. Pa's folks, five sisters and one brother, his uncle, various cousins that lived with the family—all their names were attached to the rooms they had slept in. The only heat in those rooms came through registers that went from the ceiling of the rooms downstairs and through to the floor of the rooms upstairs. Only a couple rooms upstairs were even a little warm in the winter. One was the hired man's room over the dining room. When it was time for him to get up in the morning, Mama would take the broom and tap or pound on the register to his room with the handle until she heard, "Okay, I'm up."

Above those seven bedrooms were two attics, one for the old part of the house, one for the slightly-less-old part of the house. Both were full of things that John and I could examine and wonder about and want to take downstairs. But when we asked Mama, those things always belonged to someone else—Aunt Kate or

Charlie Kinney, Gramma Fromm's bachelor brother, lived with us for many years. He didn't use swear words (at least not around us kids), but he had a lot of "by-words." Once when Uncle Charlie laid that pipe down, little Marilyn picked it up, took it behind the cookstove and tried it out. What a commotion that caused then—and chuckles for many years afterward.

Aunt Aurora or Aunt Mae. In years gone by, walnuts from our trees had been stored up there to dry. Squirrels chewed a hole into the attic and rolled the nuts around on the attic floor. If visitors were sleeping upstairs and we "forgot" to tell them about the squirrels and walnuts, they were scared witless. We thought it was fun to have squirrels up there playing, getting into Aunt Kate's stuff. Eventually the hole was patched and the races were over.

The cellar was also a wonderful place and a little scary. It had once been two cellars with two separate stairs—actually three counting the steps that went directly outside. Neither John nor I remembered when a door hole was punched through the original rock foundation and the two cellars became one. Under the old part of the house were the coal bin, the furnace, and the workbench where John and I spent many hours making tools, making Christmas presents for the cats, smelting lead, fixing stuff. We spent one winter constructing the Feef-mobile. That was an amazing car built on a coaster wagon. It had wooden sides, a wooden "convertible roof," a tin can telephone so the driver could talk to the "pusher." We added so many conveniences that we couldn't get it up the outside steps and had to take some parts off. But when summer came and we sped down the sidewalk in the Feef-mobile, we were SO proud.

The other part of the cellar had the cyclone shelter in it—a super-strong structure, about 6 by 12 feet, with a ceiling made of 4 by 10s, standing on railroad-tie legs. If a cyclone came, we knew exactly where we would hide, even if spiders and other creepy things were under there. Mama's canned tomatoes and peaches and sauces and jellies were on shelves along the walls, and big crocks held potatoes and carrots.

The story about Old Uncle Charlie and the potatoes was told and retold all our lives: One winter the folks had gotten a keg of fresh cider, and Uncle Charlie discovered that it had "turned" (become "hard"). Every-so-often he announced that he'd better go down cellar and light a candle so that the potatoes wouldn't freeze. Pa noticed that the level in the cider barrel was going down. When it was empty, Uncle Charlie didn't go down to check on the potatoes any more. Pa said, "Uncle Charlie, don't you think you should light that candle for the potatoes?"

"Oh, Con Demmit, Charlie, them potatoes can just freeze!"

The granary in winter twilight. As the oldest building on the farm, it has been repaired and repainted many times but continues to keep grain somewhat safe from mice and rats.

Granary

The granary (pronounced grain-ry) was the oldest building on the place. I liked to think that the first people who came to the farm built the granary and lived in it until the house was ready. My grandfather bought the farm from them after the Civil War. I pretended that I was one of those people and figured out where the rooms would have been. The downstairs had four nice rooms (grain bins) and a regular stairs (not a ladder) along the north wall going upstairs where there were two more rooms. If those people hadn't finished building their house before the crops were ready to be put in the bins, they had to live in and around the corn and oats.

The granary was a nice place to spend a sunny winter Saturday morning. That was where we ground feed for the chickens. Pa and John lined up the tractor, ran a belt from the tractor to the grinder and shoveled in the oats and corn for the chicken feed. I took my little iron grinder that Pa bought for me from Herman Goldstein the junk man. By hand, I ground away at a little pile of corn and oats. That sure added the finishing touch to the week's feed supply.

The granary was built on a stone foundation with the floor about two feet off the ground. That kept the corn and oats dry and gave the rats and mice a little challenge for getting a meal. It gave John and me a challenge to see if we could crawl the whole length under the granary on our bellies. The stones were solid along both long sides of the building. Another wall of stones held up the middle of the building. That left two tunnels under the granary from the north to the south side.

The space under the floor joists was only about a foot high. The dirt under there was littered with rusty tin cans and even some broken bottles. Some thoughtless people had used that place as a dump, not realizing that we would want to crawl there.

It was a challenge to our courage and strength to start at the north end and get all the way (24 feet) to the south end and then crawl the added eight feet under the little corn crib that was attached. It wasn't quite so dark under the corn crib, but it was more treacherous because there was even more junk under there.

"I'm coming through," we shouted to each other as we crawled through our own tunnel. I always suspected that John's side had just a little bit more head room than mine, so maybe it wasn't quite a fair race. But I was definitely just as strong and just as brave. Maybe even braver.

To grind oats and corn in the granary, this 1930s tractor was carefully positioned, belted and fired up to send gears spinning and ground feed pouring into buckets for the chickens and milk cows and pigs. Once-upon-a-time all that grinding had to be done by hand, so the tractor was a very welcome addition to the family.

Barn

Of course, we had the best barn in the neighborhood. We'd been in most of the other barns nearby, and we didn't mind bragging that ours was bigger and stronger and better taken care of and had more places for everything and was just the best.

Sometimes kids from Plymoth rode their bikes out to our farm on Sunday afternoons. They wanted to catch pigeons or see the new calves or pigs. When they got near the barn, they wrinkled up their noses. "Pew. It stinks in there."

"Hunh? I don't smell any stink." The barn smelled like a barn should, with all the good hay and cow and horse and pig smells. Sissy town kids!

Our barn was red with white trim around the doors and windows. It was about a hundred feet long and forty feet wide. It had been built in two stages. The back barn was the old part. Some of the supports were made from long skinny logs that still had a little bark on them. We didn't know when that part was built, whether it was already there when Grampa Fromm bought the farm back in the 1870s, or whether he built it later. Boulders formed the foundation around the outside, keeping the wooden siding and studs out of the mud. The back barn never had a floor other than dirt, as far as we knew. It was about sixty feet long and forty feet wide, divided lengthwise into three parts.

The barn at its best in about 1905. Aunt Mae is all gussied up sitting in the buggy. Max Hacker will probably be driving her to town. The photographer must have told Grampa Fromm and our Pa (as a ten-year- old) to "hold their horses!"

The middle section was a huge haymow. On either side of it were sections where the cattle or horses or pigs slept and ate. A mixture of boards and logs kept the animals out of the haymow, but allowed the cattle or horses to reach over and in for the hay. We were pretty well protected there from the half-grown calves, or the hogs rooting and squealing, or (shudder) the BULL. As soon as John and I could handle a pitchfork, we pitched hay from the haymow in the middle directly over to those boards so the animals could reach it. Usually Pa or one of the hired men like Eddie Bernhardt or Short Heidenreich was nearby to see that we didn't cause trouble with those pitchforks.

On each side, above the animals, was another floor with room for more hay. This made the pens underneath nice warm places for the stock in the winter. Sometimes straw was piled up there, and we could toss it down and spread it around for the calves and pigs to sleep on. We realized how clever our grandfather must

The barn in the 1940s. The "new" part, in the foreground, was built onto the "old" barn about 1900. A vertical white board about halfway along the side marks the joining spot. Doors and windows are used for more than entry and light—they are spaced in handy spots for manure pitching. Horses entered the wide door just behind the sheep. To the left of that is the feedway door where a terrified Deanie tried to escape from the hungry workhorse. The huge hay door under the peak would be let down by pulleys, and at haymaking time, that south side of the barn was often called "the hottest place in Cerro Gordo County."

have been because there were wooden chutes on both sides of the haymow that reached almost to the roof. When the hay was nearly that high, we could stand on top of it to pitch forkfuls into the chutes, and it landed right in front of the cattle. From the top of all that hay, we could climb onto the little platform at the north end and look out the window near the peak. If only the ladder on the wall next to the platform had gone all the way to the ground, we could have climbed up there any old time, even when the hay was all gone. But it only went about halfway down to the ground—maybe because kids like us might want to climb up there when there wasn't much hay left below, and then they might….oh, that was too scary to think about.

The cow barn, horse barn and bins for corn and oats were in the new part of the barn. It was probably added on to the old part in about 1900. Forty or fifty years later it was still "the new part of the barn." It was built on a good wooden floor, and both the cow barn and horse barn had wooden gutters in just the right places behind the horse stalls and cattle stanchions. Because we'd been around manure (pronounced muh-nerr') all our lives, we knew each kind had its own distinct shape and odor. None of them smelled particularly nasty...as long as they were not in the house. Milk cows had a fairly pleasant-smelling manure. Horse manure was rich and kind of tangy. Pig manure was hard to describe because pigs ran with all the other livestock. Worst of all was a little chicken manure stuck on the side of a shoe and carried into the house with us. Instantly Mama would shove us outside and make us wipe our shoes in the snow or grass, maybe even pump some water on them and brush them with an old scrub brush.

A cement ramp with crosswise grooves led up to the doors for the horses and cows on either side of the barn. The milk cows in the feedlot walked up that ramp, went to their own stanchions, slid their heads through the boards and started eating their ground corn. Then we went along the feedway in front of them and hooked the stanchions shut so the cows would stay put until we finished milking. The feedways in front of both the cows and horses had their own doors to the outside and doors to the oat and corn bins in the middle. When all the cows had been milked, we opened their door to the feedlot, opened the stanchions, turned the cows out, took down the manure fork from its hook behind the cows and threw all the manure onto the pile on the west side the barn. We were proud of keeping the barn nice and clean. That was especially important in fly time.

We were so glad that our barn had electric lights—even up in the haymow. On cold and snowy winter nights, we felt warm and snug under those little bulbs with cobwebs hanging around them as we pitched hay down and fed the cows and sheep. When the animals were quietly munching and bedded down for the night, we turned out all the barn lights and started for the house. It was about

Bark remains on some of the saplings that were used as supports and partitions to keep livestock and feedbunks separated.

the best time of the day—the chores were done and the lights in the kitchen windows meant Mama was making supper. We knew we'd smell the beefsteak frying as soon as we opened the back porch door.

To make our chores more interesting, we had a radio that was plugged into the light socket so it would come on when we turned the light switch on. We liked to plan evening milking time to come at the same time as our favorite programs: Lone Ranger, Hopalong Cassidy and Superman. Another favorite, Tarzan, came on a little later in the evening. His show started with a wild, jungle yell—kind of a yodel. One night at about the time for that program, Mama went into the barn, turned the light on and went into the back part where the light was pretty dim. By that time the radio had warmed up and Tarzan was coming on with his yell. She grabbed a pitchfork and ran for dear life.

The horse barn was a great place for good smells. Along the outside wall were hooks for all the harnesses and bridles and horse collars and the saddle for our pony. If we held any of them close to our noses, we could breathe in the scent of sweat and leather that we liked. The horses smelled good, too. We curried them, at least as far up as we could reach, and scratched them behind their ears. We stayed away from their huge feet with hooves as big as dinner plates.

The horse barn was well built with double-sided walls between the stalls. Near the door was a single stall where we kept our pony. In an earlier time, the riding horse had been tied there. We could only imagine what he must have been like. Beyond that single stall were three double stalls—the first for the team of bays and the second for the team of sorrels. Each stall had a hay manger in the front, next to the feedway, that we filled every night, and a feed box on each side that we filled with oats. As I was putting oats in for the bays one night, Prince got

Wooden chutes were sensibly spaced to make it easier to toss hay from the top of the haymow to the waiting livestock far below.

impatient and started to climb over the manger. I was terrible-scared and tried to get outside through the feedway door, but it was latched from the outside. I was screaming and crying and shaking when Pa came to rescue me and hear my terrifying story. The next day he took off the wooden latch—just a little block of wood screwed onto the outside of the door frame that we twisted to keep the door shut—and replaced it with a metal latch that could be opened from either the inside or outside. I felt a little safer feeding the horses after that, but I still kept an eye on Prince.

We cleaned the manure out of the gutter behind the horses every day and threw it on the pile to the east of the barn. The doors and windows in the barn were all well-placed so that it was easy to throw out manure. In the back barn the hogs and cattle were mixed together. The manure was packed in with straw and wasn't pitched out regularly. The smell made our eyes burn, especially in the spring when Pa and the hired man cleaned it out, throwing it through the windows and the doors on each side, making a long ridge of manure piles. Sometimes they got the manure spreader right under a window and pitched directly into it. Otherwise they later pitched those long piles into the manure spreader, hitched the team to it and hauled it out to the fields. We had enough cattle and horses and pigs and chickens to fertilize all the fields.

The "new" part of the barn also had a haymow over all of it. A wooden chute on each side ran from near the roof down to the feedways so the hay could easily be thrown down to the cows and horses. The chute on the cow barn side had the ladder that led from the floor all the way to the roof. The ladder was made of various boards—some smooth, some full of slivers—nailed to the studs of the wall. We knew every rung so well that we could climb it in the dark. When the haymow was totally full, we climbed to the top, crawled right next to the

The ladder made of boards nailed to the studs of the "old" barn reached from the haymow floor to the rafters. That ladder was a test of courage and eye-hand coordination for youngsters.

pigeons' nests and picked up the fat baby pigeons while their parents worried and fluttered in and out of the little hole near the peak. We could touch the track in the peak that carried bundles of hay from the huge hay door on the south clear to the little platform on the north end. By the time we were old enough to climb around in the barn, all our hay was alfalfa. It had a wonderful green smell of summer and a bouncy feel that made it good to jump in. But then the leaves fell off the alfalfa and stuck to our clothes. "Hey, you kids, quit knocking all the leaves off the hay—the cows don't want to eat bare stalks!"

In late winter most of the hay in the new part had been fed to the livestock. If we could get Pa to maneuver the carriage along the track and put it right over the division between the new part and old part, we could use the cleared floor for lots of daring challenges. We'd instruct town kids: "Grab aholt of the rope, get back aways on the floor, run and swing way out and drop into the haymow." (Some of the hay we jumped into had been in the old part of the barn quite awhile and it wasn't so precious.) Wow that was scary. And fun. We raced to get up the ladder and try it again—maybe get even farther out. Jumping off ladders into the hay was also challenging…and fun. We never broke any bones so we just kept on jumping.

The barn was a good place to find hidden nests of new kittens, nests of banty hens sitting on their eggs, messy nests of sparrows to poke down and smash, nests of mice—little pink blind creatures that we showed to our cats who devoured them. Could anyone count all the calves, pigs, lambs, kittens, pigeons, chickens, sparrows, mice, rats and other creatures that had been born in the barn? I think not!

Horses

The two work teams in the 'Forties were the bays and the sorrels. The sorrels were difficult to harness and kind of wild. The bays were good, steady horses. And huge. They always had their halters on and were stabled in the first big stall in the horsebarn. Pa harnessed them up, one at at time. The neck collars had straps that went along the horses' sides to be hitched to the whiffletree of the hayrack or corn wagon or manure spreader. Their bridles had long, long reins that reached clear to Pa where he stood on the wagon, and he drove the team out to the field.

We went with Pa any time we could. Bouncing along on the hayrack, sitting on the high seat with Pa when he was seeding oats, riding on the binder when he was cutting the oats —all of it was important work and we were helping. The horses were quiet and obedient. When Pa "picked out the headlands," (which meant walking along the edge of the corn field, picking the corn by hand and throwing it into the wagon), the horses pulled the wagon and walked along with him, needing only a "giddap" or clicking sound or "whoa" to keep them at the right pace.

We didn't ride on the horses. They were work horses and not meant for entertainment. A few times someone lifted us up to sit on a broad, sweaty back for a few minutes, but that couldn't be considered "horse-back riding." So we didn't know much about riding until Pa brought the Shetland pony home from the

Johnnie and Deanie with Prince and Barney. John was yelling, "I don't know how to park them!" Luckily the horses knew more than we did.

salebarn. "Go look what's out in the horsebarn," he said the next morning. We couldn't believe it! A pony! We named him Chief. He was beautiful—white with dark brown spots—and he was well-mannered. We rode him bare-back and fell off a lot. When we did, he stopped and waited for us to get on again. He took the bit easily and always seemed ready to go where we wanted him to go.

We rode him to the dirt road west of the farm where he shifted into a smooth gallop because the dirt was easier on his feet than the gravel. We rode him to the slough over in the northeast corner of the farm where our legs brushed through the tall slough grass as we listened to the bobolinks and redwing blackbirds. When we weren't riding him, we tied him to the light pole so he could graze and we could gaze at him. We fed him oats and hay, curried him and cut his mane so he looked smart. That first summer with Chief was great.

Winter came with short days, school days, snowy days—not good for riding our pony. We fed him and watered him and curried him but didn't ride him much. In the spring he was fat and frisky. He didn't want to take the bridle. He didn't want to go where we wanted him to. He walked slowly on the road away from the farm and then galloped flat-out all the way back to the barn and tried to scrape us off as he went through the barn door. We got a saddle, but he held his breath when we tightened the cinch, then let his breath out when we put a foot in the stirrup. The saddle slipped half-way around his belly, dumping us on the ground. And he nipped at us. We tried to take the spunk out of him, tried to wear him out with long rides. He was just naughty.

We did our chores, fed Chief and the other horses, got busy with other projects. One day we realized we hadn't seen him for awhile. "Hey, where's Chief?"

"Took him to the salebarn. You kids didn't seem interested in him anymore." It was true, and that made us feel so bad. We had neglected him. But it wasn't all our fault. He wasn't interested in us anymore either.

The team of bays, driven by one of the Bernhardt boys, pulls the manure spreader and heads out to fertilize the fields. In this way the horses are helping to clean up after themselves and also to ensure good crops of oats and hay for their winter feed.

What a wonderful pony was Chief. He taught us a lot about bridling, saddling, riding and feeding. Too late we learned that a well-fed pony was not an obedient pony.

Cows

I started milking when I was about six. Pa started me out on a gentle little Guernsey. I could only manage a few squirts at first, but before long she became "my cow" to milk. John already had two cows, and Pa and the hired man had the others. Marilyn was allergic to hay so she helped Mama in the house.

It was a good feeling to be useful in the cowbarn. I balanced on my little T-shaped milking stool, clamped the milk pail between my legs and used the "pull-squeeze" method that I'd mastered. A cow's udder is divided into quarters. The front quarters are the two teats toward the cow's head; the back quarters are toward her tail. I had my favorite quarters—one front, one back closest to me. Being kind of small, it was harder for me to reach the teats on the far side and to be sure the milk would go in the pail. Pa usually finished up for me when I started to tire out, but by then I had half a pail or so, as much as my scrawny legs could hold in one place.

Every cow had a personality. Some cows were kickers so we had to put the kicking chains on them. Some didn't let their milk down. Some swatted us in the face with a tail full of dried clods of mud. Some put a big dirty hind foot in our nearly-full milk pail. Every day was an adventure.

When a cow had a new calf, we milked out only half of her milk so the calf could have the rest. One cow was a big Holstein—really huge compared with the Guernseys and Jerseys that we usually had. She gave a lot of milk, too, and when we finished milking her, we turned her calf in with her and he got what was left. Sometimes we turned all the calves in with her, and she had four calves jockeying for positions.

When the calves were big enough to be weaned, we trained them to drink from a pail by getting them to suck on our fingers and then drawing their mouths down into the warm milk in the pail.

The little calves bawled pitifully while we were milking and they were waiting for their share. We liked them but we didn't get attached or give them names because we knew they were headed for the salebarn.

One day when the preacher and his family were at our place for Sunday dinner, we asked his daughter Frances if she wanted to see the new calf. We took her to the barn and proudly showed off the little brown calf. "Ooohhh, so cute. What's her name?" John named that calf on the spot, "Frances." The girl looked as though she didn't know if she liked sharing her name with a calf.

Our best cow was the riding cow. During the day, the cows were out in the feedlot or even out in the field. We called them at milking time, "Come boss, come booossss." Sometimes they came and sometimes the dog went after them. But the riding cow. Ahhh…. Now that was a great cow. John walked out to find her, climbed on her back and rode her to the barn. She was an all-purpose cow.

Milk cows and a few feeder cattle mill around the cowbarn door. Usually the right ones went inside to be fed and milked.

By the mid to late '40s, all the feeder cattle were Hereford steers. They had been shipped by train from Montana to Strands' stockyards in Manly, bought at the annual fall auction there and trucked to the farm in Strands' big trucks. By the next summer, when they had been fattened up on our silage and hay, they climbed back on those trucks for a trip to the packinghouse.

Steers

Steers weren't very interesting. Pa got thirty five or forty Hereford steers in the fall. They slept in the hundred-foot-long steershed, ate from the fifty foot long feed bunk in there and waded through snow and mud to the water tank down by the barn.

Morning and night, we helped feed them silage and ground corn and hay through the fall, winter and spring. We cleaned out their water tank in the summer (sometimes taking a little swim in it when the water was clean), but we sure didn't want to get involved with pitching manure out of that long shed.

In the late summer, when the steers were fat and the price was good, they were ready to go to Albert Lea or St. Paul or Chicago. Oswald Strand's big cattle trucks drove into the yard one morning. The cattle were bawling, the dog was barking, the truck drivers were shouting. We watched Johnny Sutton and the other drivers getting the steers up the cattle chute and into the trucks by poking them with the electric cattle prod. We liked to think about using that prod if we could get aholt of one.

When the last steer had been pushed into the trucks, the tailgates slammed shut and the trucks driven out of the driveway, we were done with steers for a few weeks. Thank goodness!

Pigs

My first impression of pigs was that they were mean and dirty. They were in the feedlot with the steers, and they rooted around in the dirt and manure. They ate swill and tankage and ear corn that we threw out by the shovelsful toward their trough. If it fell in the mud, they ate it anyway. When we poured swill in their trough, they fought each other for a better spot and squealed and bit ears and snouts. Mean! Swill wasn't such a great treat, I didn't think. The skim milk from the separator was poured in the barrel by the fence near their trough, and then oats and ground corn, maybe potato peelings, were put in. It soured during the day, so we didn't want to get it on our hands or clothes, but the pigs liked it. Flies did too.

Pa got twenty or thirty feeder pigs at the salebarn in the spring. They probably weighed twenty or thirty pounds, and the "boys" had to have a little surgery to change them from boars to barrows, which fattened up better. John and I didn't understand the surgery, but we were curious about what Pa and the hired man were doing in the barn with the pigs, and we wondered why the pigs were making such a racket. We climbed up into the haymow and went to the hole above the pigs' pen. Pa looked up and saw two little faces looking down. "Pa, what are you doing?"

"Uh, we're cutting out their squealers."

When the young gilt surprised us with four piglets, John's hog farming career began. From this distance, it's hard to tell which one won the blue ribbon at the North Iowa Fair.

"But Pa, they're still squealing."

"Oh yeah, there's a little bit of squeal left. We took out most of it."

After they got their strength back, the pigs were taken out to pasture for a couple months and then brought back to the hog lot to be fed until they got to be hogs—big fellas—ready to sell. At some stage between being feeder pigs and hogs, they were good to ride. Over in the feedlot west of the steershed, kind of out of sight of Mama's kitchen window, we threw out a pile of ear corn, and while the pigs were busy eating, we chose our pig and went for a short ride.

On Sunday afternoons, kids from town sometimes came out to play. Plymouth kids were used to being on farms, but Mason City kids were green—and easily impressed. Gary was from Mason City. After one Sunday at the farm, he "shared" at school, "I know a boy who can ride pigs and his name is John Fromm and that's the truth!" His teacher was Goldie Michalek, a Plymouth girl and a friend of our family. She said, "Well, I know John Fromm, so I know that's the truth."

When one of those pigs decided to become a mother, Pa made a pen in the barn for her, and the next morning there were four little pigs for us to check out. Suddenly pigs were VERY INTERESTING. John was in 4H, so they became his "project." When the North Iowa Fair opened at the end of summer, John was there with his pigs. I was home when he called to tell about the judging. He'd gotten a blue ribbon on the pen of four, and a blue ribbon on the one he chose to show alone. Standing on tiptoe to reach the telephone on the wall, I hollered into the mouthpiece, "Which one won?"

"Wha- a a t?"

"Which one won?"

"Wha a a t?"

Sheep

What made me think I wanted a lamb? I didn't know much about lambs except from reading The *Lamb on Wheels* and "Mary Had a Little Lamb." None of our neighbors had sheep. No one I even knew had sheep. And how did Pa know I wanted a lamb?

In our family we didn't beg for things or pester the folks for stuff we wanted. Many times Mama had shown us the little celluloid doll she cried for—the only thing she had ever asked her dad to buy for her. It cost a dime, and she told us how ashamed of herself she was when he sighed and told her how much that cost and how little money he had—and then he bought it for her. She could hardly even have any fun playing with it afterward. That story taught us not to beg for things, so we almost never did—we kept our wish list to ourselves.

Still, in a thoughtless moment, I must have said I wished I could have a lamb. Late one evening when I was already in my nightsuit, Pa came home from the salebarn and said, "I got you the lambs. They're out in the barn. You can see them in the morning."

Lambs? In the barn? For me? How could I sleep?! How could I wait until morning?! As soon as it was light, I raced out to the barn and … aawww. They were huge, almost as big as me. Long gray wool all over them, even on their faces. They wouldn't be any fun to play with, and they sure weren't cute and cuddly.

What could be better than spring lambs to hold and carry and play with! But look out for that ewe—she looks mighty protective.

"Pa? I thought they'd be litt-ler."

"You take care of them and maybe you'll get some litt-ler ones someday."

So I shut them in one of the horse stalls and fed them ground corn and a few forksful of hay and carried a pail of water from the pump to them and went to school. I should have said, "Thank you, Pa. I think I like them." But they didn't fit the vision that I had. And I didn't brag about them at school.

For two or three days I fed and watered them, and all they did was stare at me. Then one morning as I was pumping water into a pail for them. I heard something like a cat trying to choke out a mee-oow. I looked all around, and there under one of those sheep were two wobbly little lambs. ALL RIGHT! Now I had something to tell about at school.

And when I did chores that night, there were two more! All four lambs were sucking milk from their mothers and shaking their long tails like they were really happy. Their mothers murmured low rumbles and kept looking back at their lambs, touching them with their noses—probably to be sure that only their own lambs were sucking.

When all the lambs finished eating, I climbed over the gate and into their pen. The mothers stomped their feet at me but that was all—they didn't look dangerous, so I picked up one of the lambs. She put her little milk-covered mouth up to my ear and started nibbling. Her tightly curled wool felt nubbly under my hand and her skin was loose on her ribs. Her hooves were so new that they were still as soft as the pads on the bottoms of her feet. I looked up and saw Pa standing by the barn door, smiling. "Pa, you were right. I took care of them, and

Fifth generation on the farm, Carly and Evan Schwarz hold two new black lambs. The brick house, built in 1949, replaced the original white farmhouse.

I did get some litt-ler ones. Thank you eversomuch."

"Yup, and if you take care of these, Jeannie-with-the-light-brown-hair, maybe you'll get some more litt-ler ones next year."

Sheep were all I could talk about for days, weeks. I went on and on about how many I'd have next year and the next year...sheep all over the place, eating grass, having new lambs every spring....

John was not interested—he was into pigs, real livestock. Not sissy sheep.

But years later, long after the last cow and pig and chicken had disappeared from the farm, my sheep were still safely grazing.

Chickens

"You're flying, Old Brownie, what more could you want?" John would sing out as he ran holding Brownie over his head by her legs. Brownie squawked and flapped her wings. Maybe she didn't like flying. Mama thought Leghorns were the best chickens for laying eggs, but sometimes she got a few brown ones, maybe from a mix-up at the hatchery. They seemed more interesting to us and easier to tame, so sometimes we made pets out of them. John had that special hen, Brownie, that rode on his shoulder. Or flew.

Leghorns may have been good layers, but they were so white and so plain. And dumb. Every time I went into the chicken house to pick eggs, those old hens ran all over the place and flew and squawked as though I was some terrible menace. They didn't have any memory of ever seeing me before…or maybe they did.

At our farm we picked eggs. We didn't collect them or gather them like the city people and the people in our schoolbooks said. That was silly. Those books didn't know anything about real farms. When I was reading my lesson out loud to my teacher at our country school, I read that the farmer was explaining to the class on their field trip that a hen lays only one egg a day. I stopped and said, "But that's not true! Our hens lay four or five eggs every day and I know because I pick them." The whole school—maybe ten or twelve kids—laughed and made fun of me. Especially after school. John was ashamed of me for being so dumb. Picking eggs was probably the first chore that I got to do, and I liked it. I felt

White Leghorns were Mama's favorite chickens—good laying hens and pretty good for Sunday dinner. These hens, however, belong to Annie Pryor in Plymouth. They spent summers in her chickenhouse and yard and winters in her cellar.

important. Sometimes a hen would get "broody" and think she could stay on the nest and hatch out those four or five eggs (that weren't really all hers). I had to pull her off and pick those eggs! Sometimes she couldn't be convinced that she was not a "setting hen." That meant she had stopped laying eggs and wanted to hatch out a nest of chicks, so Mama convinced her to be chicken soup.

In a far corner, under the roost, a wayward hen or two might lay eggs that we didn't see for several days…or weeks. No telling how long they'd been there. Rather than take a chance on having rotten eggs in our pail, John and I took turns throwing them against a tree. There's nothing like the smell of a really rotten egg—it made our eyes burn.

Sometimes a hen escaped from the henhouse and laid her eggs in the barn or some other hiding place. One day she came strutting out with a flock of new chicks. We were delighted. If we managed to get her and the chicks in a little pen where they were safe from cats and rats, they survived to reach adulthood, but that didn't happen very often.

More than once I found a hen on a nest of eggs that were beginning to hatch—the chicks had started to peck holes in their shells. I decided to help the poor little things by peeling more of the shell and freeing the chicks. I was so proud of myself. But by the next day all of the chicks I had freed were dead. Huh?

Usually we got new Leghorn chicks from the hatchery. Mama came home with cardboard boxes with holes in them, and we heard the peeping as soon as she opened the car door. We were so excited when we carried the boxes to the brooder house where there was a nice place to keep them warm. A metal hood hung a few inches off the floor, and a light bulb inside it helped heat the space underneath. Sometimes the chicks pecked at each other, and what had been cute little fluffy chicks turned into ragged, bloody, homely things that we hated to look at. Mama would put a red light bulb in the socket hoping to cure them of their pecking ways, but we always felt terrible when we had to pull the dead

The chickenhouse served many purposes, only one of which was to house chickens. Sparrows also set up housekeeping above the high windows, but cats and corncobs helped discourage them.

ones out. By the time they were feathered out, Pa hitched the horses or the tractor to the brooder house (it was on skids) and pulled it to a place south of the buildings where the chicks had clean, disease-free ground to scratch in. But then they were farther away and no longer cute, so we lost interest in them.

The Claus boys gave us three little black and brown bantam chickens. They were different from the white Leghorns that stayed in the chicken house and the fenced chicken yard. Banties roamed all over, had nests in odd places, hatched out tiny, variety-colored chicks. Some of them actually avoided being eaten by cats, rats, skunks, or weasels and lived to grow up and learn to crow. "Rurr-a-rurr-a-rurrrrr." We liked them. They were clever.

John and I liked to help Mama kill chickens. We knew the expression "running around like a chicken with its head cut off," but we didn't let that happen. After Mama chopped off a chicken's head using a hatchet and a good tree stump, she

And the roof was an excellent place for beginning roof-climbers Charles and Cousin Mervin.

Deanie eventually bought five barred rock chicks with her dime, and they grew to be giant roosters. But never, ever would she mug one of them.

---

stuck the chicken in an upturned drain tile for a few minutes and then dunked it up and down in boiling hot water. Then John and I pulled out all the feathers, and that was good fun.

But the best fun was mugging chickens. First we needed to find a good mugging stick; an old hammer handle or a short piece of a hayfork handle was best. When it was almost dark and the chickens had all gone to roost in the hen house, we quietly sneaked in and found a chicken sleeping with its head hanging out over the edge of the roost just waiting to be mugged. Whack! That chicken squawked and flipped right off the roost. We could hardly keep from laughing out loud. We crept along and mugged another hen. Whack!

Probably we only stunned them, because they were usually up and around the next day. They didn't lay eggs very well after a mugging, and Mama wondered why there weren't many eggs some days. We didn't say anything.

Mean

We were kind to animals—cats, dogs, cows, even pigs. But we were mean to some people on the farm, especially ones who were smaller than we were or city kids or even some who were almost family, like the hired men.

Bob Strand was our oil man—he delivered gas and oil for our tractors—and sometimes he brought along one or two of his five daughters. We knew they had strict orders not to go in our house when Bob was filling our tank, so we tried to coax them to come in. "We've got a box upstairs with a hundred dolls. Don't you want to come play with them?" They twisted their pretty skirts or their long beautiful hair, stared at their patent leather shoes and shook their heads. We hated obedient kids. "Come on, your dad won't care. If you like one of the dolls, you can take it home." Maybe they'd glance at their dad, twist their hair harder and shake their heads again.

Did we really have a box of a hundred dolls upstairs? Well, there was a big box in one room, and it did have some dolls and parts of dolls and other toys in it, but that number one hundred was all made up.

Then there was Dixie. Her mother Margie brought her along when she came to help Mama with the ironing. We lured Dixie upstairs, got her interested in playing with paper dolls, and then one of us looked out the window and said,

"Oh, Dixie, there goes your mother, driving out of the yard. She must have forgotten you."

"Waanh! Waanh!" she cried, running down the stairs, hollering for her mommy. We loved it. Then next week we did the same thing.

Sometimes on a Sunday afternoon, a bunch of town kids walked to the farm from Plymouth—almost three miles. They played cops-and-robbers and swung on the ropes in the barn, but John liked best to hitch them up to the coaster wagon and have them pull him around the yard. He may have had a switch in his hand, too. The next week we'd see them walking up the road again. Get the harness.

Andy was our hired man. His bedroom was above the dining room, and it was heated from a register that went through the dining room ceiling into his bedroom floor. Mama woke him up in the morning by tapping on that register with a broom handle. Then he grunted, "Okay." John and I REALLY wanted to take the broom handle to his register, but we were too short. We found other ways to bother him. We opened the door to the kitchen part way, put a half-filled coffee can of water on top of the door, and waited for Andy to open the door. The can fell and spilled water all over him. Mama made us stop doing that.

Andy liked to sit in the rocking chair after supper and read our comic books. When we discovered that he couldn't read, we carefully put the comics with lots of words on top of the stack—Captain Marvel, Superwoman, even Classic Comics. We watched him shuffle through the stack to find easy ones like Donald Duck. Then we'd carry on a loud conversation about how good that Captain Marvel story was, especially when he…oh we better not spoil the ending.

Warren was the next hired man. He made pets of the calves and MY lambs—he even gave them names. He spent hours picking ticks off the lambs and squeezing grubs off the calves' backs. When we announced that some of the lambs or calves were ready to sell, he almost had a breakdown. "You aren't going to sell Mickey, are you?" He was practically crying.

"Yup, Mickey's ready to go to market," I said, "and he'll bring a good price." I probably was a little more calloused about it than necessary. "Maybe the Monday twins (he had named them Monday Morning and Monday Afternoon in honor of their birthday) will be ready to go, too." Then Warren went out to the steershed and sulked.

Every Christmas Warren ordered a crate of grapefruit from Florida, supposedly for us, but mostly because he liked Florida grapefruit. We each had a half grapefruit for breakfast and made a big deal out of eating it with our special grapefruit spoons. "Boy that was good. I think I'll have another," we said, just to watch him squirm as he calculated how many fewer that left for him.

Austin was from Missouri so he talked funny. We didn't mimic him to his face, but probably not far out of his earshot. He called "afternoon" "evening" and thought lemon pie was vinegar pie. Ugh. When Mama made pancakes for breakfast, we put the syrup in a bowl and put the tiniest spoon we could find in it for him to use to ladle it out. Then we watched as he took syrup, counting each spoonful out loud…"One, two, three…"

Dunn Gould didn't live at our house, but he came out from Plymouth to help during corn shelling or manure hauling. John and I chanted—sometimes even where he might hear us—"Gundy Gool is a fool." And we imitated the way he tromped around the place. He probably didn't even notice, but we felt sly and clever.

Uncle Johnnie was married to Pa's sister, Aunt Clara (pronounced Clarie). After they retired from their farm and moved into Plymouth, they often came out for Sunday dinner or to help out, like during threshing. Uncle Johnnie was almost blind, so we always asked him what time it was. He took out his pocket

Johnnie and Deanie might not appear to be mean, as they stand on the cistern cover, which served as home base for hide-n-seek, kick-the-can and ditch-'em, but behind those pitiful faces, devilish schemes are hatching.

watch, held it up close to his better eye, said, "Well, I swan," and put it back in his pocket. We nodded our heads and walked away. When we were by ourselves we said, "Well, I swan," and laughed like crazy. Uncle Johnnie had double-thick soles on his work shoes so that if he stepped on a board with a nail sticking up, it wouldn't go into his foot. "Uncle Johnnie, did you step on any nails today?"

When he came in for dinner, Aunt Clara said, "Johnnie, you shouldn't sit down there. You should go outside and wash up in the wash pan on the bench."

"Oh, Clarie, lee' me be," he groaned. John and I grinned at each other. There was another phrase for us to repeat when we were working around the farm!

Our sister Marilyn was ten years older than John, twelve years older than me.

During most of the time that we were growing up she was away at school. So when she came home, we didn't treat her as kindly as we could have. We treated her more like a city kid who didn't know how to live on a farm. She came with new ideas about things we should do and not do—and we thought we had been getting along just fine the way we always had.

At college she had learned about germs. "Don't eat that carrot right out of the garden. It's covered with germs." Heck, we'd wiped it off on our "overhalls," wasn't that good enough?

"Don't drink that milk fresh from the cows; it has to be pasteurized first." Then she'd boil it until there was a scum on top, and it tasted terrible. "We shouldn't be drinking water from the well down by the barn. The cow manure might be leaking in there." But that was the best tasting water on the place. Once she saw a dead branch way up in the big cottonwood tree. "That branch should be cut down before it falls on somebody." Wouldn't it be more dangerous for somebody to climb up there with a saw?

We did a number of little things to make her life uncomfortable—hiding her things, using words she didn't like, coming to the table with dirty hands. Probably the worst was lighting those illegal firecrackers under her window. Her reaction to that almost scared us into being good to her.

When she brought home that tall skinny boyfriend who tried to reform us, we were really mean. We mimicked him and mocked him and wrinkled our noses when we asked if she was going to invite Bernie next time she came home. We were proud to think that we had killed any chance of Bernie becoming our relative.

If going away to college meant that we would no longer know that the farm and the way we lived was the best thing in the world, we wanted none of it.

Milkhouse

We got up at the same time every morning, summer and winter, school days and Sundays. We had chores to do. We also went to bed at the same time every night, and sometimes we quoted Robert Louis Stevenson's poem:

> In winter I get up at night
> And go to bed by candlelight.
> In summer quite the other way,
> I have to go to bed by day.

By the age of five we had regular chores every morning and night. Our folks showed us how to do a chore—milk, feed the pigs, water the chickens—and soon we could do it by ourselves. As the seasons came around, we had other work to do like planting the garden, digging potatoes, unloading cobs and cleaning up oats and corn and soybeans that spilled under the elevator. We didn't gripe about work. We were part of the farm operation and work made us feel important. We didn't argue or complain much except to each other.

The best place to work in the summer was the milkhouse. It was always cool and clean, and it involved two of our favorite ingredients—milk and cats.

The water ran in a trough from the pump through a little hole in the north side of the milkhouse and poured into a small tank. We put the heavy two-handled

The milkhouse stands next to the windmill, which pumps cold water into a small tank inside and cools the cream in the shiny creamcans. When the empty cans come back from the creamery in Plymouth, they are washed and hung on the pole to dry and wait for the evening milking. Maybe a cat is lying under the bench waiting for a strainer pad to chew.

cans of milk there every night, and the shotgun creamcan there in the morning after separating. Another lower hole was cut in the west side of the tank and also in the milkhouse wall, so that water ran out and down a long trough to the big stock tank. That trough was a perfect stream to sail the little boats we folded from the slick pages of the Sears Roebuck catalog, or better yet, the ones we made from Mama's cucumbers. She let us have the big yellow ones that weren't good for pickles, and we cut them in half, hollowed out the centers and let them float down the trough. When they plunged off the end and fell in the stock tank, oh boy! They made a great splash, sank to the bottom and then popped up on the other side. We cheered! We were supposed to take them all out of the tank when we finished, so they wouldn't rot in the water, and sometimes we did.

Outside on the south wall of the milkhouse was a high shelf with a big copper boiler on it. When the day was hot—especially if it was a haymaking or threshing day—Pa or the hired man filled the boiler with water from the pump and put the lid on it. All day the sun beat down on that water. A hose ran from the boiler through the wall, and when the men came in from the field, they could take a hot shower inside the milkhouse. We kept a bar of soap and clean towels on a shelf nearby.

After a hot day of making hay, John sometimes got to take a shower in the milkhouse if one of the hired men like Eddie Bernhardt helped take the hose down. It was hooked on a nail too high for kids to reach, probably on purpose.

Back in the milkhouse the cold water kept running through the little tank keeping the milk and cream cold. Sometimes in the summer, Mama let us put a few bottles of Nesbitt orange pop in there, too. Nothing was any better on a hot July day than reaching way down in the tank, bringing up a cold orange pop, sitting on the cement platform around the well and slurping it down. Ahhhhh.

But work was what made that pop rewarding. We each had our jobs in the milkhouse in the morning. John's was to lift the milk cans out of the tank, pour the milk through the strainer (probably taking out all the dust and hairs) and pour it into the separator. Then he turned the handcrank that powered it, sending the cream into the cream can and skim milk into the milk can. My job was to take the disposable cotton filter out of the bottom of the metal strainer. Here's where fun with the cats came in. They were milling around, rubbing against my legs, waiting for me to toss that filter to them. It was fun to watch them growl and fight over it. Finally one dragged it away from the others, held it down with both paws, tried to chew off a chunk of the cloth and swallow it. Some cats were kind of delicate eaters and just licked the cream off the top. But some of the tough ones like Turk ate the whole thing. They ate rats and mice, bones and all, so a little fabric was nothing to them.

My other morning job, not quite so much fun, was washing the separator parts with warm soapy water. Those separator rings sometimes stuck together with a big clot of cream holding them tight. If that happened, it took strong fingernails to pry them apart. But when I scraped out that cream and flicked it in the cats' pan, they really appreciated it.

When John learned about electric motors and figured out how to put a belt on one so it turned the separator crank, he cut his workload in half. But no electricity helped me wash, pry, brush, rinse, dry, carry and dump my workload. Sometimes I wisht I was a boy.

Backhouse

The backhouse stood near the garage and had its own little sidewalk leading up to the door. It was near enough to the house to be convenient, but far enough away to keep odors from drifting into the kitchen. Mama kept it nice and clean—she painted the bench around the toilet holes a dark green enamel—shiny and smooth with no slivers to hurt us—and she scrubbed it with a pail of water from the washing machine after the clothes were washed and hung out on the line.

There were three holes to choose from, each with nicely rounded and sanded edges. On the right was a little hole about six inches across but made even smaller with two boards nailed along each side from underneath so no little tot would be afraid of falling through. In the middle was the most popular hole, a middle sized one, and on the left was one we thought was a little bigger.

We liked to use the backhouse in the warmer months and in the daytime (we had indoor plumbing for winter and nights) because we didn't have to take off our muddy shoes or risk going in the house and being asked to sweep the sidewalk or take potato peelings out to the chickens. Generally John and I went to the backhouse together because it was a good place to speculate and to make plans. I sat on the little seat and John sat in the middle—the best place because he could look straight out the door and comment on which cats or dogs or chickens were walking past.

The backhouse in winter was not as popular as in other seasons, but even then it was more convenient than the indoor facility for kids with dirty overshoes or snowy woolen mackinaws.

Three holes presented some choices. The small hole on the right has lost one of the boards nailed to the underside which would keep a little kid from falling through—or from worrying about it. However, worrying about spiders would still be in order.

We never closed the door when we were in there because there was a hook on the outside, and what if someone came along and hooked it while we were inside! We speculated on who might have used the third seat because it wasn't nearly as worn as the other two—maybe someone really fat, like Old Uncle Charlie who used to live with us. We speculated on spiders that might live under our seats and bite us. I was terrible-scared of spiders. We speculated about what we would do if we got to be a nun and changed our minds and wanted to get out. Since we always saw nuns in pairs (in their long black gowns shopping in the dime store), we decided we would distract the nun assigned to us, buy regular clothes, get a locker in the bus depot and hide the clothes in that. One day we'd slip away from the others, go to the bus depot, change our clothes and take a bus out of town. We spent lots of time planning this getaway, adding details. What about a wig?

We planned our day's work while sitting there—maybe hunting sparrows or digging angleworms or trying to find that new litter of kittens in the barn.

Grown-ups might have called it play, but with John anything we did was always called work. His often-quoted complaint about me was, "When I want her to wook, she won't wook, and when I don't want her to wook, she wooks."

Store-bought toilet paper was for the indoor toilet. For the backhouse, we had a wooden box filled with *Sears Roebuck* and *Montgomery Ward* catalogs and maybe some *Wallace's Farmer* or *Capper's Weekly* magazines. They provided reading and discussion material for John and me, and then they were "recycled."

The index pages were thinner and softer, so they were the first to be torn out and used. Once in a great while Mama decided she wasn't going to need one of her sewing patterns any more, so the backhouse inherited that soft paper, too. Best of all were peach papers from the crate or two of peaches that Mama bought and canned every summer. Peaches were each wrapped in soft pink squares that we flattened and took like prized possessions to the box in the backhouse. We really liked peach season.

Our backhouse didn't have a crescent moon cut into the side like those pictures we saw in the funny papers. Ours had two little windows way up high on either side. One glass didn't fit very well, so when the wind blew, it rattled back and forth and could have been scary. That's why it was good for both of us to be there together.

Unlike the funny papers and other things we read, ours was never called one of those fancy names like "outhouse" or "privy." It was called what it was: a backhouse.

Smokehouse

A strange little building sat in a neglected part of the yard out behind our woodhouse. It was unusual because it had a cement floor, no windows and a galvanized pipe poking through the roof. When we went inside, we couldn't really see much, so that made it a little scary. Spiders were sure in there, but where? I hardly ever went in the smokehouse because of the spider thing and because nothing of interest was in there, except a few old pieces of furniture.

But we liked to stand in the doorway just to smell the inside. When we opened the door, ham and bacon came to our noses. Not in either of our memories had anyone smoked meat in there, but the wooden walls remembered.

We imagined that Uncle Charlie Kinney was in charge of that. He probably had time to cure the ham and bacon, to keep the fire going, to hang the meat on the hooks that were still on the rafters and to keep the critters away from it. Gee, the farm must have been an interesting place way back when Pa was a kid.

Playhouse

John and I mostly did what we called *work*. So why was there a *playhouse* at the farm? It was probably named that when it was built one summer as a birthday present for Marilyn. I was born when she was twelve, so the folks reasoned that both girls would get a lot of use out of a playhouse. She later admitted that she was almost too old to play in it herself, but she knew I would spend a lot of time there. And she was right.

The playhouse may have been about eight by ten feet and had a steep, steep roof that allowed us to make an upstairs by putting boards across the joists at the roofline. I could stand in the center of the upstairs and reach my hand way up and still not touch the peak. But of course I didn't do that because there were spiders up in that peak.

Still, instead of *playing*, I usually worked in the playhouse. On those first spring days, after the playhouse had been closed up all winter, I worked at cleaning everything. I moved all the furniture outside—both of the little cupboards that Gib Peshak had made for Marilyn, the drop-leaf table and two chairs, the little round kerosene stove, the little sink that had running water (if I put water in the reservoir at the back), the two rocking chairs, the stepstool we used so we could get upstairs, the orange crates that I pretended were beds, even the curtains from all three windows.

Marilyn and Cousin Norma Lou Haegg from Cedar Rapids have propped their dolls in the window box of the playhouse, and the girls stand nicely like the well-mannered young ladies that they are. But they might not realize that in the peak of the playhouse roof, which does not show in this photo, "up there be spiders."

When the playhouse was swept and scrubbed, the spider webs destroyed, and all the furniture back in place, I put the blankets and pillows on the orange-crate-bed and stretched out, exhausted. Maybe tomorrow I would play there with my dolls.

But I was never very good at playing with dolls. Mama told me that Marilyn dressed hers every morning and put them to bed every night, but I didn't live up to that example. Instead, I brought the cats into the playhouse, tucked them into their beds in the upstairs and tried to keep them from leaping down and escaping. I liked best to bring in a mother cat and a whole nest of kittens. I wrapped them up in doll blankets and rocked them—sometimes they really went to sleep, unlike dolls that only shut their eyes.

Playhouse furniture was made for Marilyn in the 'Thirties and lovingly used by children and grandchildren for many decades. Here Julia Schwarz prepares for a tea party—yummmm, delicious red tea!

The playhouse had many other uses. It was the robbers' den when we played cops-'n-robbers. It could also be the bank with cupboards filled with paper money and tinfoil coins—loot for the robbers. It was a great place to hide when we played hide-'n-seek. If I got way back against the wall upstairs and covered up with a blanket, nobody could find me.

In the winter Mama sometimes used it as a place to hide Christmas presents. Once she hid a little red wheelbarrow she had gotten for John, but she forgot about it until my annual spring-cleaning. "Hey what's this doing in the upstairs?" After that we always expected to find forgotten Christmas presents up there, but we never did again.

Except for times when the Smith kids or the O'Hern girls came over for a game of ditch-'em or hide-'n-seek, we didn't really *play*. We mostly thought that what we were doing was working around the farm. But we did have fun, and the playhouse was a good and useful part of that fun. And we were never ever bored.

Haymaking

Bringing in any of the crops—oats, soybeans, corn—was important work for John and me. We always found something "helpful" that we could do that was also interesting. Like most farmers in our area, Pa made hay three or four times every summer, so John and I had lots of chances to learn how to help with haymaking.

The hay was really alfalfa, a modern crop in those days. Alfalfa seeds were mixed with oats. Early in the spring, as soon as the fields dried enough for the horses and wagon to cross, those seeds were broadcast from the seeder at the back of our straight-sided wagon. We liked to help Pa seed oats on those first warm days in April because it was good to be out in the field again. Sometimes John got to sit on the wagon seat and drive the horses, even though they didn't need much driving. They generally walked at just the right speed straight across the field, and Pa turned them around. I helped Pa shovel oats into the hopper with my little shovel.

During the summer the oats grew tall and green, with tops that filled with grain ("headed out" we said), then turned yellow and dry. Without our noticing, the alfalfa was growing slowly under the oats. Later in the fall—after the oats had been cut, bundled, shocked and threshed—the field turned green with the young alfalfa, now getting its chance at full sunshine. Next summer it would be ready for haymaking.

The horses are ready, the hayrack is ready and Short Heidenreich is ready to slap the reins and head for the hayfield. Even after Short took over his father-in-law's farm a mile down the road, he often came back to help with haying.

The next June, when about one-fourth of the alfalfa had little purple blossoms, Pa hitched the horses to the mower, with its long, scary sickle-bar, and started out to the hayfield. I liked to ride along, holding on to the back of Pa's seat. If the mower scared out a nest of little rabbits, I thought I could capture one and have it for a pet. Never-mind that everyone told me wild rabbits would die if we tried to keep them in a cage. Still, I liked to watch for rabbits and to watch the alfalfa tremble and fall in a nice swath behind the mower.

The hay dried in those long swaths for a couple days, if we were lucky and no rain came. Then Pa hitched the horses to the hay rake and headed to the field again. The rake was set at an angle—the right side was nearer the horses and the left side angled farther back—so that the tines picked up the hay, fluffed it, rolled it together and dropped it in long windrows to the left of the rake. We rode with Pa for hours watching the rake pick up and deliver nice long rolls of hay, listening to the meadow larks and bobolinks sing from the fence posts,

The hayloader was cleverly designed to roll over the windrow of hay and pick it up with its "teeth." At the Pioneer Museum in Mason City, Charles Butler, Marilyn's husband, recalls using one like this on the Alabama farm where he grew up.

watching our dog Rex run alongside the horses or tear off across the field after a rabbit. What could be better than that?

Unless it was loading the hay onto the hayracks. That happened a day or two later when the hay in the windrows was nice and dry and no rain was on the way. For this the horses were hitched to the hayrack, a huge (to us) flatbed on wheels, with slats on the sides to hold in the hay. The front of the hayrack had slats that went up higher, and in the middle of the front slats was something like a ladder where the driver stood, holding the horses' reins. The hayloader was hitched behind the hayrack.

The hayloader was another amazing invention. It was as wide as the hayrack and almost like a conveyer belt on wheels. It slanted up from the ground and over the back of the hayrack. It had tines that picked up the hay, moved it up the slanting bed and dropped the hay into the hayrack. John and I hitched our coaster wagon behind the hayloader and rode out to the hayfield—quite a procession: horses, hayrack with Pa standing on the front driving the horses and the hired man (probably Short Heidenreich) standing next to him, the hayloader, John and me jolting along in the coaster wagon at the end.

When we got to the first windrow, Pa and Short laid out the hay slings on the floor of the hayrack. These were ropes that carried the hay up into the haymow, and they had to be laid out so that they would hold a huge bundle of hay together and so that the ends could be easily found when we got back to the barn. Usually the men tied the ends to the front and back of the hayrack. Then the horses started down the first windrow and the hay started moving up the hayloader and the men started pitching the hay evenly around the hayrack

Later in the 'Forties, haymaking changed from loose hay to baled hay and from horses to tractors. Short liked to drive the Farmall "M" because it was full of horsepower, but its fuel couldn't be grown on the farm.

and John and I started picking up stray wisps of hay to put in our wagon to take back to our own little barn.

When the hayrack was half full, Pa said, "Whoa" to the horses. Then he and Short laid out another set of hay slings so another bundle of hay could be loaded on top of the first. Pa said, "Giddap," slapped the reins, and we were off again. When the hay was piled as high as the loader could push it and as the men could pitch it—an impossibly huge load that looked really scary to me– we headed for the barn, leaving the hayloader in the field for the next load. If plenty of help was available, another team of men and horses and another hayrack was already in the field to bring in another load while we put the first load in the barn.

At the barn, the big haymow door was open and the carrier on the track in the roofpeak was at the very south end, ready to pull up the first bundle of hay.

The peak of the barn roof still holds the track that carried a sling-load of hay a hundred feet to the far north end of the barn.

We heard the men saying that the hottest place in Cerro Gordo County was right there. Our horses knew exactly how to take the hayrack to the right spot by the barn, and John and I watched as the men pulled the hayslings around the top bundle and hooked them to the ropes leading up to the carrier. By this time, Johnny Sutton, with the long trip-rope in his hand, had climbed to the little platform inside the barn at the north end.

A second rope led from the carrier all the way along the track, then down to the floor and around a pulley on the northeast corner of the barn. There, a helper hitched the rope to another team of horses or maybe a tractor. When everything was in place, Pa shouted and that team or tractor started pulling the rope that lifted the hay bundle up and into the barn. John and I took up posts as the relayers of the shouted orders: "Hip!" Hup!" "Ho-oh!"

When Johnny Sutton, on that platform with the trip-rope in his hand, reckoned that the bundle of hay had traveled along the track to the right spot, he pulled the trip-rope, dropping the hay in exactly the right place. It was an art.

He shouted "Ho-oh!" for the team to stop pulling, John and I relayed the shout (louder and maybe more times than necessary), and then we unhooked the rope and dragged it back to the barn, ready for the next bundle to be lifted into the haymow. We liked to be so helpful.

Between loads, someone (not us kids) had to pitch the hay around to even it out all across the haymow. That was a terrible job, dusty and itchy and hot. We felt sorry for the man who got that job.

Oh we thought it was great when the barn was full of fresh hay! It smelled so green and lush, and we knew how the cattle and horses would love it in the winter. We didn't know about spontaneous combustion. We didn't know why our folks kept a nervous eye on the barn for many days after the hay had been piled deep in the haymow.

Threshing

The absolutely most exciting time on our farm was threshing time. (We pronounced it "thrashing.") But getting the oats ready for threshing required a lot of steps. First Pa hitched the horses to the binder and rode on it out to the oatfield. I liked to hang on to the back of Pa's seat and watch that magical machine work and listen to all the sounds around it. The horses' big feet made nice crunches in the fresh oat stubble. I liked the smooth, slicing sounds of the sickle-bar cutting the oats and the whirring sounds of the reel as it pushed the falling oats onto the wide canvas belt. I liked the clattering sounds of that canvas slapping against the rollers that kept it moving, and I liked the squeaks and rattles of the gears and chains as they worked. I especially liked the swish as twine was wrapped around each bundle (which was about as big as Pa's two hands would reach around), and the clicks as the machine cleverly cut and tied the twine in a neat knot. There was a zing as two prong "fingers" kicked the bundle out onto the tines of the cradle or bundle carrier. I watched Pa count the bundles in that cradle, and when there were enough for one shock, he pulled a lever and the cradle dropped those bundles onto the ground. I was fascinated by that binder—it knew so much. After a round or two I was tired and went back to the house. Before nightfall the oatfield was dotted with little piles of bundles.

A day or so later the shockers came—two or three men who went from farm to farm shocking oats. They moved through our field of bundles and turned

This photo was taken from *The Wonder Book of Knowledge* by Henry Chase, but in the 'Forties that would have been Pa on the seat and Prince and Barney providing the horsepower.

it into a field of neat little tents of oats, which helped the heads dry. Shocking was a tough job. John and I tried to do it, but our shocks were sloppy and soon fell over. We were supposed to grab the twine of two bundles (one in each hand), jam those two down into the stubble at an angle so the tops were touching, pick up two more and jam them into the stubble making kind of a square tent. Then we did the same with more bundles to make a nice tight shock. But, of course, ours were neither neat nor tight.

When the shocks were dry, it was time for threshing. We belonged to the neighborhood threshing ring with four or five other farmers. All of them got together and decided whose farm to go to first. Then the big wooden Altman-Taylor threshing machine (the "thrasher"), parked between two of our corncribs during the rest of the year, was rolled up the road to the lucky first farm. All the other neighbors with their teams and hayracks and tractors and able-bodied youngsters headed to that farm as soon as their morning chores were done.

Late one afternoon we'd hear a commotion up the road and see a tractor pulling the threshing machine to our place. John and I started jumping up and down, running around the yard, holding our breath as we watched that big outfit make the turn into our driveway and to the chosen spot west of the steershed. We climbed all over it, inspecting every part as though we'd never seen it before, even though it had been standing by the corncribs all year. "Tomorrow!" we'd shout.

Tomorrow came—with all the men and horses and hayracks and tractors. On the huge thresher, every gear and sprocket had to be greased, every belt had to be tightened, every spout had to be adjusted, and then Art Rose hollered, "Stand back, boys, I'm gonna twist her tail." The tractor that powered the threshing machine's main belt started up, the threshing machine shuddered, the gears and chains and belts squealed, and we were in business.

Early that morning Mama was busy peeling potatoes, chopping cabbage, boiling eggs. Getting dinner for threshers was the main event of the year for Mama. She had been making plans for it for weeks. She had dug potatoes and carrots in the garden, killed and dressed chickens, canned rhubarb sauce and bought extra cans of coffee and baked beans. A day or so before the threshers came to our farm, Mama started baking pies and cakes and rolls. She thawed out beef and pork roasts from the locker. She always hoped that threshers wouldn't be at our place on a Friday because it was so hard to make a good meal of fried fish or canned salmon. We were about the only Methodists in our neighborhood, with Catholic families all around us. We all respected the practice of no-meat-on-Fridays.

Usually Marilyn (if she was home) and Violet Foster or Violet Wasicek or Violet Gedville would be there to help Mama get dinner, but we had our jobs to do, too. "Run down to the cellar and bring up a couple jars of peaches. Take these peelings out to the chickens. Go down to the well and fill these thermos jugs with cold water for the men." As soon as we could, John and I offered to

Threshing time was exciting and fun for kids—lots of noise, lots of neighbors, lots of horses and tractors and lots of food.

put the jugs of cold water in our coaster wagon and take them out to the men so we could escape from the house to be where the action was.

By then several teams and hayracks had been to the field, and the men—walking alongside the racks—had pitched on as many bundles as the racks could hold. A parade of teams waited for their turn to move close to the noisy machine where the men pitched the bundles into "the monster's gaping yaw." Oats poured out of one spout into a waiting grain wagon, and straw flew out of the big spout to make the straw stack. Someone like Johnny Sutton steered that spout, pulling ropes to aim the straw evenly on the ground for a good, round, solid stack—one that cattle and pigs could chew on around the edges or find a nice spot to sleep in, but one that would not fall over and smother them. The straw felt so slick in our hands and smelled warm and dusty. It would make great bedding for the livestock in the barn and steershed in the winter.

John and I liked to stand on the tires of the grain wagon and watch the oats pour in. We always picked out the plumpest grains, pulled the husks off and

Like a giant mouth, the threshing machine's self-feeder would gobble up oat bundles as fast as the men could pitch them. This metal machine retired next to Highway S56 near Bolan, Iowa. Our wood-bodied thresher had a shorter life.

The Farmall "M" that powered the thresher in the 'Forties still purrs for Marguerite, William and Sophie Schwarz in 2010.

chewed the oats. When they were still a little soft, they were so good to eat raw. I usually stayed near Pa wherever he was working, but John was more adventurous. I would see him climbing on hayracks, sitting on idle tractors, checking out the straw pile. "Oh, John, be careful."

Meanwhile, up at the house, Mama worked over the hot cookstove, putting in another stick of firewood, turning the roasts around in the oven, taking out the tins of hot bread. What delicious aromas came from each pan! About that time we showed up in the kitchen. Mama gave us orders to set the table in the dining room and get the wash pans and soap and hot water and towels ready to put on the bench outside near the back door. We'd much rather be helping Pa out by the threshing machine, but we'd been sent to the house.

At noon all the machines were shut off and the dozen or so men headed in for dinner. They were laughing and joking and telling stories while they washed

"Thrasher at rest."

up and came into the house. The Lava soap had removed some of the grime and sweat from their hands and faces, but their workshirts and overalls, soaked with fresh sweat and dusted with oat chaff, brought a not-unpleasant smell with them. These neighbor men were comrads who had worked hard all morning on what could be dangerous machinery. They were a team and each one had special jobs. They depended on each other.

When we carried out the big bowls of mashed potatoes and gravy and meat and string beans and corn and coleslaw and pickles and bread, everybody quieted down and got to work doing some serious eating. John and I ate in the kitchen with Marilyn and the Violets. By the time Mama brought out the pies, the conversation started up again. I was always in the dining room by then so I could listen. Sometimes the stories were scary—about accidents on threshers or with dynamite or about cyclones that ripped through other farms. "Drove straw right into the telephone poles. Took the baby right out of the buggy. They found the baby a half mile away, poor little thing." Those stories fed my nightmares for months to come.

Depending on the weather, threshing at our place took three or four days—days of hot, sweaty work for the men and long hours of cooking and washing dishes for Mama. If each of the other farms in our threshing ring took that many days, threshing could take the better part of a month. Our farm was usually the last so that the threshing machine could be put back in its place between the corncribs.

When only one bundle was left on the last hayrack, Short picked it up on his pitchfork and tossed it into the threshing machine saying, "There's the bundle we've been looking for all day." Then the tractors shut down, the rumbles and squeals of the threshing machine wound down, the neighbors and horses and hayracks clattered out of the driveway, and everything was strangely quiet. We looked at each other, tired and a little bit sad. Threshing was over for another year.

Silo

At supper one night, Pa said, "The men will come tomorrow to start building the silo." The silo?! John and I looked at each other. We could already see ourselves climbing a silo. "Where will it be? What will it be made out of? Will it have that nice checkerboard on top?"

"Nope," Pa said. "No checkerboard."

"Why not?" I whined. "I want that pretty white checkerboard like on Frank Montgomery's silo."

"Because if we need to make it taller in a few years, the checkerboard would be in the middle, like on that place up by Manly, and that looks silly," Pa said. That seemed like a good enough reason. We sure wouldn't want a silly-looking silo.

Pa was a progressive farmer. He took over the farm at eighteen when his father, the Civil War veteran, died. He left high school (where he met Mama) before he graduated. His uncles and neighbors helped out, gave advice, but he was always reading and asking questions and keeping up with new ideas. He wasn't satisfied just doing things in the same old way. Now silage was the coming thing.

Running a hand over those "ever-so-slightly curved tile" proves that they are just as rough as when they were put in place in 1946.

The next day the men arrived with a truck full of reddish-brown tile, a cement mixer, ladders, scaffolding—the whole works. We looked the tile over carefully. It was rough on the outside and smooth on the inside and looked perfectly straight. How could they make a round silo with straight tile? The men patiently showed us that it was curved ever-so-slightly—this was going to be a big silo, twenty feet across—and that slight curve was plenty to make it come out round. Then Pa told us to stay back and not ask any more questions. We watched as the silo grew, row on row of slightly-curved tile. Quite a few feet up, the men started putting in some iron rungs a couple rows apart. A ladder! Great! "But Pa," I asked, "How come it starts so far up? We'd have to put a ladder up there to climb it."

"Right. That's why I'm keeping the ladders away from you two."

I was especially excited when I saw one of the men brushing white paint on the rough side of some tile. When we came home from school one day, there was the finished silo, with that nice white checkerboard at the top.

And just in time. Silo filling season was starting! We found out that making silage was as much fun as threshing oats or making hay. It took a whole new kind of machinery—a chopper (pulled by the "M," our Farmall tractor), big wagons to hold the chopped green corn stalks and ears, a silo filler with a long pipe (wired to the iron ladder) that sent the silage way up over the top and into the silo. The sound of that filler was music to our ears.

Zinng. Zinng. And the groan of the tractor that powered every forkful upward. UrrrUmmm. UrrrUmmm.

We liked the smell of fresh silage, all chopped and milky. Sometimes we chewed on the little slices of corn-on-the-cob. The white-faced feeder cattle liked the fresh silage, too, but they liked it even better when it had soured, like sauerkraut, in the next few months. We didn't eat it then, but it still smelled good.

The ladder on the outside was only one way to get to the top of the silo. The other way was inside the steershed, hanging on to the silage doors that would be removed one by one as the silage was thrown down from the top. The metal chute kept the silage going straight down to the floor in the steershed, and it also kept me from being scared as I climbed way up to the top. There, hanging on to the top bars, I could look out over our fields, over the west road, over Smith's fields, over to the cement plant in Mason City, maybe even over to Clear Lake.

It was worth every minute of terror climbing up that high just to see so far. But it wasn't something I needed to do very often.

Iron rungs make a sturdy ladder so the stout-hearted can climb to the top of the silo. But it's not for sissies.

Windmill

Naturally we had a windmill, (we pronounced it "win-mell.") Every farm had a windmill over the well, which was usually by the barn because a lot of water was needed to fill the stock tank for all the cattle and horses and pigs. Eddie Bernhardt's folks had a windmill right by the house. We thought that was swell because they didn't have to carry pails of good drinking water from the barn clear up to the house every day like we did.

Our windmill was really, really tall, lots taller than the barn, maybe taller than the cottonwood trees out in the grove. On one corner were L-shaped rods that made steps clear to the top. (I think Pa had battered down the first few steps so we wouldn't climb them. Ha! He could have saved the effort.) We never climbed very high, but every so often Toad Sutton climbed to the top to grease the gears for the wheel. There was a little wooden platform at the top where Toad could balance with his grease gun. He was used to climbing because he was an electrician. He could climb those electric line poles with nothing but spiked boots and a leather strap to hold him onto the pole.

Toad fascinated us because he knew a lot about things we were interested in. He could find Indian arrowheads and other artifacts and had a big collection. I asked him how he knew so much about where Indians lived and hunted in years gone by. He said he tried to think like an Indian and find a place where there was a river or spring with a high hill or bluff nearby so he could spot

The windmill towered over the barn, the milkhouse, the trees and especially us. We were scared to even think of climbing to the top.

FUN ON THE FARM IN THE 'FORTIES

a deer or buffalo. Kids who wanted to hunt for arrowheads with him usually found some that they could show off at school. Toad also knew where the best wildflowers grew, and he transplanted some to his garden. He even had lady slippers there, but he never told us exactly where he found them.

When Howard McNitt's baby boy was about two weeks old, we went over to see him. Old John McNitt, the baby's grandfather, said, "I guess we'll have to take the S-T-E-P-S off the windmill." We could hardly wait to get out to the car to burst out laughing—just think of that little bitty baby knowing what STEPS were even if old John hadn't spelled the word, or to think of that baby crawling out to the barn and climbing the windmill! Crazy!

Seems there was almost always enough wind to fill the tank for the cattle, but once in a while we had to take out the pin that held the rod (the pitman) on the pump to the shaft that ran up to the wheel, put the pin into the handle, and use our muscles to pump enough water to keep the cows from bawling. We had to push pretty hard on the handle to push it down, but the sucker-pipe helped lift it up. Sometimes the handle came up a little too fast, like the time it flew up and hit John in the mouth, chipping his best front tooth.

The well had been hand-dug many years earlier. We were told that it was thirty feet deep. Once when Pa and the hired man were replacing the wooden platform that held the pump, we peered into the well and saw the stones laid up along the sides. We tried to imagine someone digging down and down, putting rocks around the sides as he dug, until finally his feet reached water. What if it had filled up too fast? How would he get out? Would he float to the top? Would somebody grab him and pull him out?

The windmill had a gearshift that controlled the speed of pumping. If there was only a slight wind, we squeezed the two parts of the gearshift's handle together and put the gear in the highest notch. The wheel caught all the breeze and pumped water slowly. When the wind was strong, we put the shift at a lower notch. When the tank was full, we put the shift all the way down. If we didn't shut it off in time, the tank overflowed, causing a big mudhole all around the tank. The pigs liked that, but the cattle hated it.

Sometimes we forgot to shut the windmill off at night, especially if there was no wind all day and the tank was almost empty. Then late in the evening we heard that screech and whine as the wheel caught the wind and started pumping water, and we felt glad that we wouldn't have to pump by hand the next day.

Sometimes a storm came up in the night and the windmill pumped for all it was worth. Then Pa had to pull on his "overhalls" and overshoes and go take care of the windmill.

We spent a lot of time taking care of the windmill. Sometimes we had to cut short our Sunday drive to take care of it. Sometimes we didn't even have time to go to Birdsall's for a chocolate chip ice cream cone. Sometimes we had to interrupt our "work" to shut it off or turn it on. But it spent a lot of time taking care of us, too. Somebody really smart must have invented it.

Trees

Dozens and dozens of trees were on our farm—in the grove west of the barn, in the woods north of the house, along the fences, around the buildings. We liked our trees and all the different things they were good for. But some of them scared us, too.

Grampa Fromm had planted cottonwoods about eighty years earlier and they were huge. We looked at that big cottonwood east of the house when a storm was coming up and the wind was blowing from the east. We tried to calculate how tall it was, and whether it would hit our bedroom if it fell. Somehow we learned that when the sun was in the right place, we could measure the tree's shadow and compare it to our own shadow. If my shadow was three feet long and I was four feet tall, and the tree's shadow was a hundred feet long, that meant the tree was…oh gosh, the top branches would hit our bedroom! And the tree was probably still growing, so if it blew down next year….

Those big cottonwoods on the south edge of the grove were not quite as scary. If they blew down, they would only hit the cornfield or other trees. Still, if the wind was blowing, we didn't go out to the grove that much.

Unless it was maple syrup time. When we found that strangely carved stick and asked Mama what it was, she told us about Uncle Charlie tapping the maple trees out in the grove and bringing in the sap to be boiled into syrup

The row of cottonwoods in the grove might appear to be one tree. But this becomes six towering giants from a closer vantage point.

Encircling the rough-barked trunks requires three people holding hands.

---

and sugar. Right then we began hatching "the maple syrup plan." We spent that winter whittling and drilling the tapping pegs, finding enough old coffee jars, putting wires around their rims and making wire loops to hang them on the pegs. When the snow began to melt a little bit around the base of the trees, it was time to tap. We set off for the grove with our sled full of jars, pegs, drill and hammer. How thoughtful it was of Grampa to plant maple trees!

Other trees on the farm had been planted, too—some pines, the silver maple that Pa planted for Mama, maybe the butternut and some of the walnut trees. We "stepped the husks off" the walnuts every fall by putting on our oldest overshoes and crushing the green husks and kicking the walnuts around on the ground. The husks would leave a brown stain on our hands or shoes or pants' legs that was almost impossible to get off. After the nuts dried enough so that they wouldn't stain our hands, we picked them up and took them down cellar or up to the attic. Next winter when Mama asked us to crack some walnuts and she would make fudge, we were glad to take our hammers and crack them open on a little chunk of iron rail from a railroad track. Butternuts were easier to crack open, and they were especially good in cookies.

One year Pa planted a row of Chinese elm trees as a windbreak north of the barn. We felt pretty fancy having trees we imagined had come from China. They were just spindly little things, but they grew quickly and within a few years they really did become a good windbreak.

In the spring when the plum trees were ready to bloom, we spent a lot of time around the plum thicket. Plums had the prettiest white blossoms and the sweetest smell, so we sometimes broke off a little branch of buds to take in the house to "force" it to bloom early. Plum blossoms and the violets around the plum thicket were usually perfect on May first. When we made May baskets to take to the neighbor kids, we always put a few plum blossoms and violets in next to the popcorn and candy corn. Later in the summer we picked the ripe plums, and Mama made plum jelly that she poured in jelly jars and sealed with paraffin. We sometimes sneaked a little warm paraffin and molded it or chewed it. Tasteless stuff but it felt interesting in our mouths.

About the time the plum blossoms were fading, our apple trees started to bloom. Even though the limbs were prickly, they didn't keep me from climbing up into the tree to be right next to all those sweet pink-and-white blossoms. Usually a robin scolded me for getting too close to her nest. She just made it more tempting—I had to see if there were any of those pretty blue eggs in it. I never fell out of a tree, but I came close more than once. By threshing time the apples were red-and-yellow-ripe. If Pa left an empty hayrack under the apple tree, we could pick buckets full for some of Mama's apple pies. Her pies with that tender flaky crust and those warm, sweet, sliced-up apples were just the best!

Most of our trees were box elders that grew wherever they wanted. They leaned at odd angles, and that made them good for climbing. We liked them for that. Before we got in the car for a Sunday afternoon ride, Mama would go out to one of the box elder trees and break off a switch to take along—a nice long one that could reach anywhere in the back seat. We didn't like them so much for that.

Box elder trees were soft trees that fell down a lot, dropped branches and rotted in the middle—they were generally second-class trees. But in one way they were our best trees—on a cold winter night, their wood in the cookstove kept us nice and warm.

Box elder trees grew at odd angles, rotted easily and dropped branches constantly. But they were fun to climb.

Snow

Oh boy! We loved snow! When we looked out the kitchen window toward the barn and saw what had happened overnight, we worked fast to get on our woolen snow pants, our four-buckle overshoes, our woolen coats buttoned up to our chins, our caps with ear flappers and our warm woolen mittens.

Snow was great. We could do so many things with it that we argued about what to do first. Usually the first thing was to measure it. If the snow was light and fluffy, we waded through it to the barn or the woodhouse and showed how it came up to our ankles or knees or, better yet, waists. If enough wind blew overnight, the snow was packed in hard drifts, and we walked on top of them. "Looky, I'm taller than the pump." "Looky, I'm taller than the milkhouse." Sometimes, if the wind had been really good, the drifts went right up to the edge of the chickenhouse roof. Then we climbed the drifts and walked around on the roof. We were giants.

It was good to be on top of the snow, but it was even better to be under it. The best place for tunnels was in the ditch across the road north of the house. Maybe we used shovels or buckets, but mostly we used our hands. John started tunneling in one place, and I started two or three yards away. The plan was to dig down and straight in, and then curve the tunnels so they met in the middle of the ditch. We carried out a lot of snow and packed some on the sides

Blasting down the hill, skidding between trees, John makes a triumphant run.

and bottom of our tunnels and kept going in the right direction by hollering a lot. "Where are you now?" "Are you turning yet?"

Except for all the hollering, it was quiet in the tunnel. The snow muffled the sounds, and our bodies shut out most of the light from the open end, so it was a little bit scary, too. But pretty soon we had a break-through. I kept digging and John's black mitten poked right through the snow. Yaaay! A little more digging to make the final passageway big enough, and then we crawled all the way through. For some reason I usually made the first pass. John encouraged me by hollering, "Are you making it?" "Are you at the turn yet?" And I hollered, "I'm coming through!"

It's a perfect day for sledding, and Rex is happy to be in on the action.

One good set of tunnels was not enough. We had to branch out and make tunnels up and down the ditch. Sometimes our reckoning was wrong, and we dug and dug but never connected our two tunnels. Sometimes we dug too close to the top. Then we poked our heads up through the snow and yelled, "Cave-in!"

After a good morning of digging, we showed up at the back door, and Mama met us with her broom. She swept the snow off us from top to bottom—with maybe a few extra swats on our bottoms. It was great to finish a hard morning of tunneling with a good sweeping off!

We had a fine hill north of the house for sledding. The snow on it had to be packed down by making quite a few "runs" first, but then we picked up our two-runner sleds, got back aways, got up our speed, ran with our sleds, flopped down on them at just the right time, grabbed onto the steering handles, raced down between two trees and skidded around just before we hit the third tree. That was a perfect run. Sometimes we hit a tree or two and had to do it over until we got it right.

John thought the hill would be better with ice on it, so he splashed a few buckets of water on the trail. After that froze, we grabbed our sleds and ran and flopped on them like always. But the first time down was so terrifying that even John was afraid to try again. Everything we knew about steering was of no use. The sleds were out of control! So we shoveled snow on the path and went back to sledding the old-fashioned way.

It was never too cold or snowy or windy for us to go out and do chores. Pa always did a few chores before breakfast and checked the thermometer on the north side of the woodhouse before he came in. Sometimes he reported,

"Twenty five below," so Mama folded a dishtowel cornerwise and tied it over our mouths and noses. Then out we went to do our chores.

Snow, blowing snow, blizzards were all just normal parts of winter and not something we worried about. If the roads didn't get plowed for a week or so, that was okay. We had a coal bin full of coal, a woodhouse full of wood and cobs, a good furnace in the cellar, a good cookstove in the kitchen, plenty of jars of Mama's good home-canned food in the fruit cellar, flour and sugar in their bins, fresh eggs and milk and meat for us, plenty of hay and corn and oats for the livestock. "Let 'er snow, boys."

School

My first day at country school was awful. I wasn't really going to school yet. I was only visiting for an afternoon in the spring before I would start the next fall. I was lying on the floor near the big south windows drawing a picture when I heard a commotion nearby. I got up and then slipped in something and fell down again. Something messy was all over my long brown stockings. The big boys were laughing.

"Shirley spilled her custard pie and Deanie fell in it." They thought it was so funny!

Miss Scholl helped me take off my stockings and wash my hands and legs. Then she said I could go to the playground and swing. I sat in the swing and looked up at the big windows. I knew I had done something I shouldn't have. When school was over, Mama came in the car to get John and me. Miss Scholl handed her a paper bag and said something in a low voice. On the way home, John told Mama that Shirley threw up on the floor and I fell in it. Oh, so that's what was on my stockings. I was afraid to eat custard pie for a long time after that.

When school started in the fall, I was a full-time student in "primer." (We didn't know the word "kindergarten.") We probably had a dozen students, one or two in each grade, but I was the only one in primer. I had a tiny notebook

On the first day of school, lunch buckets in hand, Johnnie and Deanie walk down the driveway on the first leg of the mile-and-a-quarter route to school. In Iowa, country schools were spaced so that no student would have to walk more than two miles one way.

in which I wrote each new word as I learned it. John was in second grade, and Miss Scholl had written on his report card "Inclined toward mischief." This was mentioned from time to time at the supper table.

We learned a lot in school. We younger kids sat in the little desks by the south windows. While we did our arithmetic, we listened to the big kids talk about history and geography. Boy, someday WE'D be in those big desks and already know all that stuff about oceans and mountains. If we didn't forget it by then. Reading and spelling caused me some problems. I knew how to say a lot of words like batayta and cubberd, so when some strange words appeared in schoolbooks, words like potato and cupboard, I was confused. Didn't those

Michael Smith rode his pony to school while his twin, Patricia, held the bridle. Johnnie is third from the left.

people know how to spell? And when the picture showed yellow hay in the barn, I told Miss Scholl that was wrong. Hay is green. "No," she said, "alfalfa is green. The old-fashioned hay is yellow." I still thought I was right.

We learned the best music from *The Golden Book of Favorite Songs* and sang to the accompaniment of the Victrola. Sometimes the teacher picked out the tune on the piano. "Play with both hands. Play with both hands," we begged. And at recess we played the best games—pom pom pull-away, too late for supper, fox and geese (in the fresh snow) and softball, using the work-up rules because we never had enough kids for two teams.

Every day John and I discussed how to get to school. We talked it over during chores or at breakfast and considered whether we would walk or ride our bikes. We thought about riding our pony Chief, tying him to a fencepost in the school yard all day, but we never did. Other kids might bother him or beg us for rides. Besides, he didn't like to have us ride double on him, so one of us would have to walk anyway.

Usually Lime Creek #6 had a dozen or so students. Some must have been playing hide-n-seek when this photo was taken of Roger Davison, Donald Petznick, Deanie Fromm, Johnnie Fromm, and (back row) Mary Agnes O'Hern, Mrs. Seaman and Lester Rowe.

"All-ee, all-ee, all in free!" Hide-'n-seek is over and fourteen students are present and accounted for. Front row: Tommy Smith, Bobby Nehring, Deanie Fromm peeking around unknown boy, Donald Petznick, Donovan Crooks, Darrell Crooks, Johnnie Fromm. Row 2: Mary Agnes O'Hern, Audrey Snyder, unknown brother and sister, Calvin Wendell behind Loretta O'Hern and Mrs. Seaman.

It was a mile and a quarter to school. During good weather we always got there on our own. We usually met up with Mary Agnes and Loretta O'Hern at the corner and walked the rest of the way together. Sometimes in really bad weather, Mama took us in the car, but only if she was going into Mason City anyway. Once when we walked to the corner, the O'Hern girls' aunt was giving them a ride, and she invited us to ride with them. I said no. "My mother told me not to ride with strangers." Oh, how I was teased about that. It was their aunt, for goodness sakes.

The War was almost always on while we were in country school. John started in 1940 and I in 1942. On Fridays we brought our dime to buy a War Stamp. When we filled a whole stamp book, we turned it in for a War Bond. If we heard a plane go over, we all ran to the windows and looked up to see if it had a big star on the underwing. Yup, one of ours.

Brother Charles and his taller friend perch on a sign nailed to the flagpole by the big south windows of our country school. The basement could be used for recess on very rainy days, but most students would rather play outside in any weather.

Deciding how we would get to school was usually a matter to be discussed during breakfast. Sometimes we rode bikes, sometimes we walked, sometimes we rode with Mama, but only if she was going past the school anyway.

Some of the older kids knew the different kinds of planes. B-29 was the big one with four engines. If we saw one of those, we felt safe for some reason. Our men were up there protecting us. If a German plane flew over our little school in the middle of the country, our guys were there to shoot it down.

Fall, winter and spring, we almost always walked home with the O'Hern girls. Sometimes we hung our lunch pails on a long stick. Two people carried it for a while and the other two ran around exploring. We almost always argued, too. Farmall tractors were better than Allis Chalmers. Ford cars were better than Plymouths. Catholic church was better than Methodist—Catholic church had been started by Jesus, but Methodist church was just started by a man. We asked Mama about that when we got home.

And when we did get home, we were starved. We always made ourselves a bread-and-butter-and-peanut-butter sandwich, listened to our two favorite radio serials, ("The Sparrow and the Hawk" and "Cimarron Tavern."), changed clothes and went out to do chores. Homework? Naw. With one teacher and a dozen students, country school allowed plenty of time in the day to do all the school work. Homework was farmwork.

The last day of school was usually on John's birthday, May 15. Sometimes we went to Coopers' Woods and looked for wild flowers. We found buttercups and trilliums and Dutchmen's britches and hepatica and violets (violets weren't so unusual because we had them in our own woods). If we were really lucky we found Jack-in-the-pulpits. Mama knew the names of all of them. The plum trees were in bloom and the little creek that ran through the woods was rippling, so Coopers' Woods was the prettiest, sweetest-smelling, nicest-sounding place for an end-of-school/ birthday celebration.

Sometimes we went to East Park in Mason City to have a last-day-of-school picnic. We swung on the really high swings and slid down the really tall scary slide and threw up and then went home. School was over for another year. Y-e-e-a-a-a-y!

Church

We went to Church-and-Sunday-School in Plymouth every Sunday morning. Sometimes we went to church on Sunday evenings, too, either at our church or at the Free Methodist church a block away from our church—but a long way off in style. They had "altar call" and "testimonies" and other personal things that made us nervous. Sometimes we went to "revival meetings" in a tent set up on neutral ground. That really made us nervous. People cried and shouted, "Praise the Lord" and "Save me, Jesus." We preferred our quiet Methodists.

Every Sunday we liked to paste a star next to our names on the wall chart at Sunday School. When we had a row of five regular stars, we got to paste on a gold star, and when we had enough gold stars, we got a little pin with "Methodist Sunday School" printed on it. The pin had a place for attachments—little bars that stood for a year of perfect attendance. Some of the big kids had five or six bars attached to their pins. We coveted those pins, which was a problem. Our teachers told us not to covet anything that was our neighbor's.

We also learned that God knows everything we do. If we are good, we will go to heaven, but if we do too many bad things ("sins," my teacher, Mae Peshak, called them), we will go to hell. When I asked Mae how many bad things it would take to send a person to hell, she didn't have a definite answer, so I wrote a letter to God: Dear God, How many bad things have I done and how many more will I do for I go to heaven or hell?

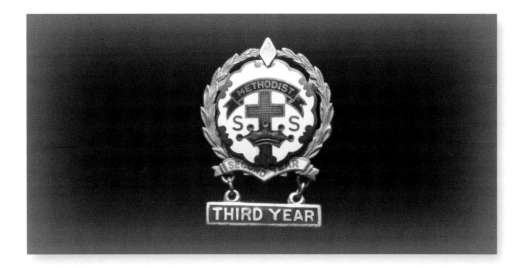

The coveted Sunday School pin.

Left: Dressed up and ready for church, Mama would have studied her Sunday School lesson and done her hair on Saturday night. On Sunday mornings she put the roast in the oven for dinner and could look forward to a nap in the afternoon.

I wasn't quite sure how to mail the letter, so I left it on my little desk for awhile. Marilyn found it—she was much more reverent than I—and was carried away with what she thought was my plea for forgiveness and my promise to be good from now on. She wanted to talk to me about my desire for salvation. I squirmed away. I didn't tell her that wasn't what I was asking God. I just wanted to find out what my score was and how many more bad things I could do and still go to heaven. I had probably done plenty already, like picking that flower from the neighbor's bleeding heart and putting it in my pocket. When Mama found it, she talked to me about stealing and made me throw it away. I never-ever went near another bleeding heart plant!

I couldn't remember all the bad things I had done, but I knew God was writing them all down in that big book. I would have to answer for each of them someday. We weren't lucky like the Catholic kids. They got their sins wiped away every week at confession, but we Methodists hung on to ours until doomsday.

Plymouth Methodist Church was like a home away from home. Church-and-Sunday-School, Bible School, choir, MYF, church suppers and bazaars, the weddings and baptisms and funerals of all our family and friends—we were always there. Our longtime friend, Joan Strand Rich, played the organ for most of those events.

Town

We didn't go places in the car much during the 'Forties because gas was rationed and tires were impossible to replace. During the war, the speed limit in Iowa was 35 in the country, which seemed pretty fast to us—we couldn't remember ever going any faster than that. The roads near the farm were filled with mudholes in the spring and snowdrifts in the winter. We had plenty of mud and snow, but almost everything else was rationed. Mama carefully saved "ration coupons" so she could buy necessities like sugar and coffee and shoes. During the Great Depression our family didn't buy much because we didn't have enough money. During the War we had a little more money—prices for farm products were good—but we didn't have enough coupons.

Plymouth (population 300) was about three miles from our farm, and we usually went there a couple times a week—once for Church-and-Sunday-School and once to take cream to the creamery and get butter in return. We liked to watch Mick the buttermaker test our cream by dipping a little spoon in it and tasting it. We watched his face to see what he was thinking. Then he spit that cream toward the floor drain and made a checkmark on our cream ticket—we always hoped he marked it "sweet" not "sour."

Then we went to the locker next to the creamery and picked up some of our pork and beef stored in our rented freezer space there. We had to remember

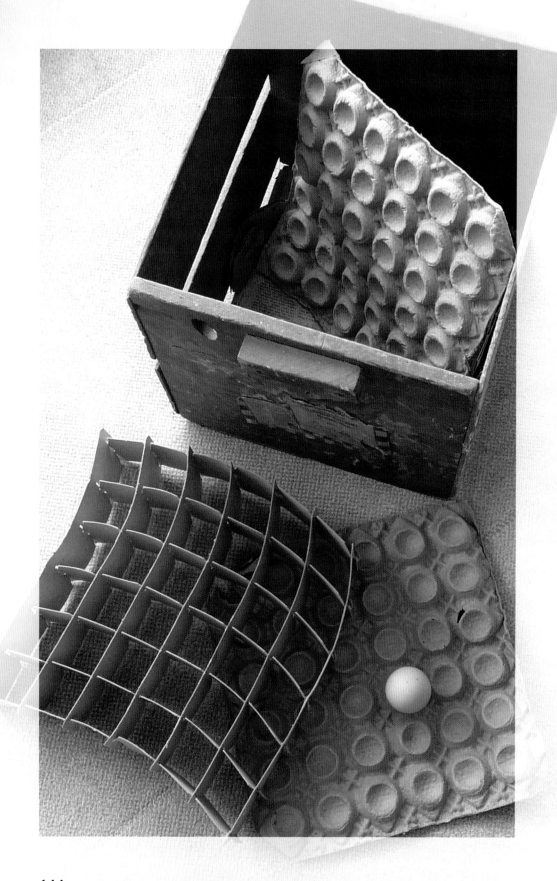

144 PACKIN' CATS FOR THE ARRR-MEEE

Going to town meant getting the eggs ready to take to Barrett Brothers Grocery to "trade." All week the eggs had been picked and taken down to the cool cellar where we sanded off any dirty spots and put them in the cardboard dividers in the wooden egg crates. But carefully carrying the crates of twelve dozen eggs up the cellar stairs and out to the car was not a job for a kid.

---

to take our locker key from home (Number 288) for our freezer and to bring a coat because that hall leading to the freezers was ever-so-cold.

Plymouth had three grocery stores and we gave a little business to each of them. Each store had a specialty: lunch meat and dried beef in one, ice cream in another, bananas in a third.

Graversens' hardware was our favorite. It had a smell that we recognized as soon as we stepped in the door. The smell of ropes, harness leather, dynamite and popcorn (they always had a chamber pot full of fresh popcorn) blended together to give that unmistakable hardware-store-aroma. Local men sat near the cash register on tractor seats mounted on nail kegs. Sometimes we got to sit there, too, and listen to their stories. There was always a lot of laughing and sometimes, after a sideways glance in our direction, a low mumbling that we couldn't quite make out. Open bins held nails of all sizes, bolts and burrs, screws, hooks. Oh, it was fun to run our hands into a bin full of washers that felt like coins. What we wouldn't do if we had a whole bagful of them!

We went seven miles to Mason City (population 25,000), the Cerro Gordo County seat, every Saturday to take piano lessons from Mrs. Patchen and to trade eggs for groceries at Barrett Brothers Grocery Store. Mason was a wonderful city because it had so many good stores along the main street (Federal) with so many good windows for window shopping. They were all close to Barretts. Sometimes if John and I got some extra change to spend, we bought something to eat at Dad's Hamburgers in the alley near Barretts. A hamburger was a dime and a bottle of orange pop was a nickel. We liked to watch the two

Plymouth on a busy day: the telephone office is humming, Peshak's grocery store is "trading," the barbershop is buzzing and Valley's café is serving hamburgers and orange pop.

Right: Federal Avenue, Mason City, Iowa, in the 1940s. We really liked Merkels Department Store, especially when Marilyn was working there, because it made us feel important. Photo courtesy of the Mason City Public Library Archives.

sisters-who-never-smiled make the hamburgers, pressing them down on the griddle with a trowel-like tool. The sizzle and smell was a treat to ears, eyes and noses. We went to the dime store to see what we could buy with a dime. Maybe a goldfish, maybe a pad of paper with lots of different colors, maybe a rubber ball, maybe candy.

If Mama was shopping in Damon's Department Store, John and I liked to go downstairs to the restrooms where we got a little paper cup with a pointed bottom. Then we went halfway up the steps to the second floor, aimed the empty cup at a spot just behind a clerk standing at the first floor counter, and dropped it. She would jump and look around when it made a loud "pop." By then we were out of sight, holding our sides to keep from laughing.

The hatchery where Mama got her baby chicks was down the street from the dime stores. Once when I was looking in the newspaper, I saw the hatchery ad that said "Day-old chicks, 2 cents each." Wow. I schemed how I could get my own chicks. The next Saturday when we went to Mason City, I took my dime allowance (a penny for each day I made my bed, a penny for each day I didn't

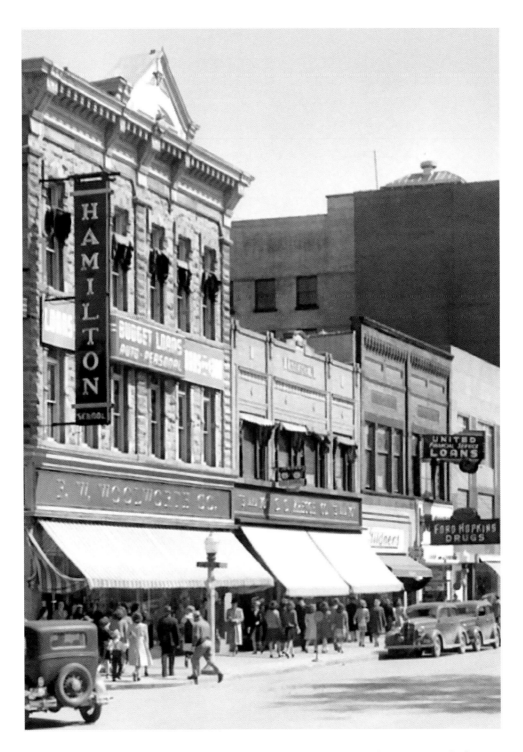

The Woolworth and Kresge dime stores were the best places to be on Saturday mornings—looking, planning, wishing, scheming and dreaming—and then spending our weekly dime allowance. Photo courtesy of the Mason City Public Library Archives.

Everything we could want was within a block or two of this corner. In Damon's we could ride the elevator or scare the clerks, in Osco Drugstore we could examine the candy counter or get our Brownie camera film developed, and in Curries Hardware we could check out the fishpoles and BB guns. Photo courtesy of the Mason City Public Library Archives.

get in trouble at school) and instead of going to the dime store, I walked the three extra blocks to the hatchery without telling anyone. I thought of all the things that could go wrong. Maybe they wouldn't sell just five chicks or maybe they wouldn't sell any to a kid. Maybe they wouldn't have any more two-cent chicks left. I hadn't reckoned on the hatchery being closed, but it was. Feeling very glum and disappointed, I walked back to the grocery store where Mama was still shopping. My dream of having five little chicks of my own had fizzled.

Almost always we went to Mason on Saturday mornings, but the absolutely best time to be there was at night in the summer when all the street lights were on and all the store windows were lit up. It was a whole different city then. Cars drove slowly down the street and the policeman stood in the middle of the intersection with his whistle. He blew "tweee EET" and motioned the north-south bound cars to go. Then he held up his hand, blew "TWEE eet" and motioned the east-west bound cars to go. We watched him through many "twee EETs."

And we heard the "zink birds." They only came out at night and they only lived in Mason City, no place else. They darted down from the sky with a loud WHURRR and then flew around the streetlights saying, "zink, zink."

Couldn't nothin' be better on a Saturday night than eating a chocolate chip ice cream cone from Birdsall's, watching the policeman direct traffic and listening to the zink birds.

Neighbors

Living on a quiet gravel road with neighbors a half-mile away meant that we paid attention to lots of little sounds. We could hear a car's tires crunching on gravel long before the car was near our place. It was easy to tell if it was slowing down enough to turn in at our driveway, and John and I liked to see neighbors, but we were pretty shy around strangers. Sometimes we would hide.

We stayed home almost all the time, but learned a little about the world outside the farm from people who stopped by and from weekly trips to town. Neighbors came to talk about crops and weather and the war. Hardly a day went by that somebody didn't stop in.

Some of our neighbors had been our hired men and lived with us when they were young and single—Eddie Bernhardt, Johnnie Sutton, Short Heidenreich. When the war came, they went into the army. We were happy that Short was gone only a couple months. When the army sent him home because of his flat feet, he gladly came back to live with us again. Soon he married Alice and they made a little apartment out of our parlor. It was fun to have a young couple living there. They even invited me to dinner with them sometimes. After a while they took over her father's farm a mile away, but Short still stopped by often. We liked him because he joked a lot. Every spring he said, "Spring is sprung, the grass has riz. I wonder where the flowers is." We always laughed.

We heard that Eddie's plane was shot down and crashed in Sweden—a neutral country. He was okay and was working for a Swedish farm family. When we looked at a map, we couldn't imagine Eddie being in Sweden. After the war he married Violet, and they lived on a farm just down the road. He still came to our place and helped with field work and helped John and me make kites and fly them. When he and Violet had a baby boy, I was beside myself with excitement. I LOVED babies but hardly ever got to hold one or play with one. I could ride my bike down to their place every day!

Then something strange happened. Pa and Eddie had "a falling out." I knew something was wrong, but never asked Pa, of course. John hung around the men out in the field and heard that it had something to do with Eddie not picking our beans when Pa wanted him to. Our two families didn't get together anymore. I didn't get to play with baby David anymore. One day they moved to a farm several miles away. Nothing like that had ever happened before—our friends and neighbors were always our friends and neighbors.

Door-to-door salesmen were regular callers at our place. Mr. Clark was the bread man who had a whole bakery in his truck—sweet rolls and doughnuts and all kinds of bread and buns. Sometimes he gave us a doughnut. The Watkins man brought the flavorings Mama used when she baked—vanilla, maple, cinnamon, nutmeg—and the pine-scented "Healum Grease" that Mama put on our scratches and cuts and scrapes.

Pa liked to play jokes on the salesmen. Once he was in the yard when the Watkins salesman drove in and asked if the wife was home. "Yes, she's in the house, but she's real deaf so you'll have to talk loud to her." The salesman came out of the house in a few minutes and drove away. Pa went in and asked Mama, "Did you buy anything from the Watkins man?"

"Who'd buy anything from a guy like that? Comes right up and shouts in a person's face!"

One regular caller was Herman Goldstein who came to pick up scrap metal—like the tin cans we had flattened—and to bring news and give advice. If we listened closely, we could learn a lot. He had a daughter who went to New York City on the train. Imagine that!

Some folks came to do special work at the farm. Chris Carstens was a tiler. He hand-dug the tile ditches and laid the orange drainage tiles that carried rainwater from our fields to the slough. From there it ran to the creek (pronounced "crick") and then to the Shell Rock River. The ditches were about five feet deep and as wide as a man. When we walked out to take coffee or a sandwich to Chris, we saw only his head and shoulders and a tiling spade full of dirt flying out of the ditch. He was from Denmark so we liked to listen to the way he talked. We looked into the ditch and saw that water was seeping in, making the bottom muddy. Chris was barefoot to save his shoes from the mud. How could he dig? "I yoost hook dat big toe over da shuffle und poosh it inta da groun."

Chris came up to the house at noon for dinner, and we listened to his stories about coming to America and then going back overseas to fight in World War I. He rode a bicycle carrying messages from the front line back to his unit. Some of the other soldiers thought he was a spy. They were sure of it when payday came and he didn't get any pay. Or any mail. He got along by gambling. After the war, his papers were straightened out and he got all his mail and back pay. So he became a farmer—and a tiler. Once when Chris had a sore shoulder, he told us that he went to see Doc Chilson about it. Doc Chilson said, "Well Chris, you're in a bad way."

"Oh? Iss dot so?" Chris said.

"Yes. I've seen a lot of fellows die from what you've got." Doc didn't have much of a bedside manner.

One of our favorite visitors was Jim Otzen, our insurance agent. When he told stories, he laughed so hard his little belly shook. One story was about Chris Tietz and the bees. Jim kept bees in his back yard in Manly and was always on the lookout for a free swarm of honey bees. Chris Tietz lived in an old country schoolhouse and had a lot of kids. He raised pigs and sold hog feed, kind of a farm-to-farm salesman. He had a bunch of hogs in hoghouses with pens around them. Tietz had found a swarm of honey bees in one of those houses. Tietz told Jim that if he wanted to, he could come get that swarm. When Jim got there, the kids—10 or 12 of them—were all sitting on the big steps of the schoolhouse. He brought out his smoker and bee veil, all the equipment. "Oh you don't need all that," Tietz said. "I'll go out and get them in this old nail keg."

"Okay, if that's what you want to do, go ahead," Jim said. So Tietz headed into the hog house and then Jim heard the darndest beller you ever heard and here came Tietz running and leaping over the hog fences with the swarm of bees right after him. He came toward the steps of the house where the kids were sitting. His wife, who was more than portly, saw what was happening and she yelled, "Run, kids. Run for your life! Run for your life!" The kids hopped off those steps just like rabbits and ran in all directions. Tietz charged up the steps and right into the house and jumped into the bathtub.

Oh, we loved Jim's stories.

War

John was six-and-a-half in December 1941. He clearly remembers the beginning of World War II. We came home from church, and as usual Mama was getting dinner ready to put on the table while Pa went out to check on the cattle and the water tank. And as usual Mama turned on the radio. After listening for a few minutes, she told John, "Run out and tell Pa to come in. There's something on the radio that he needs to hear."

We spent the rest of the day listening to the radio and playing quietly. Nobody talked much. Something pretty scary was happening in a place none of us had heard of, Pearl Harbor.

Marilyn was a student in Mason City Junior College. Within a few weeks it was almost like a girls' school. The few boys still in college were classified 4-F. They had been rejected by the armed forces because of physical problems.

Pretty soon there were a lot of songs on the radio that were related to the war. Marilyn played the piano, so she bought sheet music for some of the popular tunes and played and sang, "There'll be blue birds over the white cliffs of Dover," "I'll be seeing you in all the old familiar places," "The last time I saw Paris, her heart was warm and gay," and "I'm dreaming of a white Christmas just like the ones I used to know." Sometimes we sang along and added a little extra to some of the tunes. "I'm dreaming of a black Christmas, just like the

Eddie Bernhardt was our hired man both before and after the war. His plane was shot down and he spent many months on a farm in Sweden. We really liked him because he treated us kids like real people.

coal bin down below." With a serious face, Marilyn tried to impress on us that our soldiers were over there in the Pacific where it was hot and dry and full of bombs and guns and fire. We should think about how sad it was that they couldn't be home at Christmas with their families and snow and presents. So we sang our versions of the songs in the barn—Marilyn didn't go there.

Farmers supported the war effort in many ways—one was growing hemp to be made into rope. Pa decided to have a small field of hemp, which meant a new process of planting and harvesting and hauling to the new hemp plant by Mason City. Hemp grew to be really tall, but it was lightweight. When it was cut and piled onto the hayracks, the men could stack it way-way up. Sometimes some of it slid off before the hayrack even got out on the road. Speed Platts was a daredevil guy who came to help us at harvest time. He said he could pull one of those hayracks to the hemp plant with his motorcycle. The motorcycle didn't always start so well, and it died pretty often, so Speed had to lay it over on its side to get it started again. When we saw him finally drive out on the road pulling that huge load of hemp, we just shook our heads and wondered if he would ever make it. The next morning the empty hayrack was back at our place. And Speed was probably bragging all over town.

Airplanes went over our place quite often during the War, maybe two or three times a week. They always sounded scary to me because I knew that some planes dropped bombs. I was really worried if I heard one at night—I'd read in my Weekly Reader about those poor little kids in England who had to sleep in bomb shelters during the air raids at night. I worried that I might have to sleep in a strange underground tunnel.

One cloudy, rainy morning Mama was washing clothes and hanging them on clotheslines strung in the dining room (Monday was washday rain or shine), when we heard a plane go over the house really low. "That plane's in trouble," Pa said. A few minutes later our telephone gave one loooong ring—the line ring. Whenever something important happened, the telephone operator in Plymouth opened all the lines and made an announcement. This time she said, "An airplane has crashed two miles south of Plymouth."

We got in the car and drove toward the big black cloud of smoke. Lots of other cars were parked along the road near the cloud, so we stopped and asked questions. "What happened? What was it? Was anyone killed?"

It was an army plane. Must have had engine trouble. Two or three soldiers were on it, all dead. I stayed by the car with Mama, but John ran up as close as he could to the plane and picked up a little piece of its metal before they made him go back.

After that, whenever the morning was dark and cloudy and rainy, we thought about the sound of that plane and the black cloud of smoke.

PLYMOUTH HONOR ROLL				
Frances Chilson	Donald Willford	Harold Sponheim	John Sutton	Arleigh S. Lunde
	Alva Pearce	Everett H. Vikturek	James Woodhouse	Joe Chehock
	Forest Lane	Gordon Hogan	Duward Willford	Glen Lane
	Howard Hedegaard	Theodore Peshak	Norman Heinselman	Harold Borchardt
	Leland Faktor	Arnold K. Hanson	Gerald Borchardt	Raymond Reynolds
	Harry Ehlers	Marlin Kinney	Reuben Hovel	Gene D. Reynolds
	Harold Pedelty	Clyde Staley	Wayne Rezab	Edwin Kugler
	Robert Willford	Dwayne Hedegard	Orville Miller	Ivan Lantz
	Clifford Halbrook	Fred Bernhardt	Earl Sprung	Cleo Heinselman
	Gust Bernhardt	Walter Pearce	Kenneth Sprung	Cecil Vickerman
	Harold Buell	Lester Jirsa	Henry Bernhardt	Edward Hovel
	Jack Jirsa	William Morford	Reuben Arzberger	Donald Tapscott
	Hulbert Lyke	Leo Hovel	Ralph Peterson	Lloyd Woodhouse
	William Ingersoll	Joe Sura	Eddie Bernhardt	Ralph Trego
	Wayne F. Ingersoll	Frank Barta	Leroy Strand	H Wilbur Hill
	Lester Chehock	Elmer Navratil	Leroy Helmer	Earl Hugi
	William Reynolds	Leland K. Snell	Eldred Harmon	Dr. Kenneth Prescott
	Robert Urbatsch	Lawrence Carmany	Ellsworth Helm	Gerald R. Prescott
	Ole Framstad	Delmar Casper	Dale Helm	P. Richard Young
	G. Page Reynolds	Leon Wise	Norman Rezab	Louis J. Bartusek
	Gene Davis			

The white placard stood next to the Plymouth Town Hall and recorded the names of all the people from the area who were serving in World War II. Gold stars appeared next to those killed in action.

We talked about the draft a lot, who was being called up, who was going for their physical (whatever that was). Once Pa took us to Manly where we drove past a house that had yellow paint splashed all over the front porch and up the sides. Chicken feathers stuck to the paint. Pa told us that the guy who lived there had been drafted but refused to go into the service, so people did this to his house. It meant he was yellow, he was a chicken. We knew what those words meant. We had used them a lot with kids in the schoolyard. Now we knew how serious they were. Would we ever use them again without thinking about that house?

In Plymouth a big white billboard next to the town hall had the name of each serviceman painted on it in black letters. Some names had a gold star next to them—they had been killed. We knew some of them: Leland Faktor who flew with Jimmy Dolittle and crashed in China; Lester Chehock; Gust Bernhardt (Eddie's brother); and Harry Paxton (Mama's cousin) who died in Italy. Sometimes we saw gold stars in the windows of houses in Mason City.

Pa's big radio brought us War News and serious, worried looks to our folks' faces. But it also brought us Jack Benny, Edgar Bergen and Charlie McCarthy, Red Skelton, "It Pays to Be Ignorant" and many hours of laughter. Eventually the old radio retired to a shelf in the garage where cats pay it no attention.

Someone in the family had died in the war. John and I nudged each other and pointed, but we knew better than to talk about anyone who died, especially if Pa was nearby. Talk of death put all kinds of sadness on him, and we didn't want to see him being sad.

We forgot that unspoken rule one day in April 1945, when our favorite after-school program "Cimarron Tavern" was interrupted to announce that President Franklin Delano Roosevelt had died. We were mad that we couldn't hear the rest of our program. We ran out to the steershed, where Pa and Cliff Marrow, the hired man, were pitching out manure. We told Pa that the

FUN ON THE FARM IN THE 'FORTIES **159**

President died and some other guy was going to be President. I said I thought it was Dewey—the only other name I knew because of the election the year before. "No, it wouldn't be Dewey," Pa said, "maybe Truman." Yup, we agreed, it was probably Truman.

On a sunny afternoon in August, John and I were down by the barn when Hazel Marrow, Cliff's wife, drove into the yard, jumped out of the car and hollered, "The war is over!"

Pa and Cliff came out of the machine shed, Mama came out of the garden. "What? What?" Japan surrendered. Something about a really big bomb. John and I hardly knew what to think. The war had been with us during almost all of our memory. What would life be like now?

Waste-Not

"Use it Up, Wear it Out, Make it Do, or Do Without!"

Decades before "Reduce, Re-use, Recycle" became a widely-used slogan, we were practicing it every day on our farm. Almost nothing was thrown away, at least not until it had been used over and over again.

Plastic was unknown in the 1940s, and paper bags were seldom used in the stores we went to. Because we grew almost everything we ate, we bought very few groceries. We took our eggs in a specially made wooden crate to Barrett Brothers Grocery Store in Mason City and traded eggs for the few staples we needed. Generally we brought those groceries home in the egg crate. Sometimes the clerks put the groceries in a smaller cardboard box from their back room. Once in a great while they put them in a wooden orange crate or peach crate. That was a reason to celebrate!

A wooden crate could be used for so many things. It could be a bookcase upstairs or a set of shelves for Mama's homemade soap or fruit jars or for some tools down in the cellar. Cardboard boxes were also prized—for storing old clothes, old dishes or broken things waiting to be fixed.

A few of the groceries we bought came in cans—coffee, tea, baked beans, salmon and (for special occasions) fruit cocktail. When the cans were empty,

John and I washed them, cut out both ends—that was tough with that old stubby can opener—and took them out to the sidewalk by the garage. There we flattened them with hammers, pushed the lids inside, and saved them for Herman Goldstein, the junk man, when he came to pick up scrap metal for the war effort.

Flour and sugar came in cloth bags. When the bags were empty, we pulled the threads to open them into one flat piece, and Mama hemmed them for dishtowels. She also made "under-pillow-cases" to put over the pillow ticking before putting on the nice embroidered pillowcases. That helped keep the feathers in and our greasy hairs out. Some flourmills used bags with pretty prints on them, and some people made those into shirts or skirts or dresses or aprons. Mama saved the nicest ones, but when she asked me if I wanted her to make one into a dress for me, I shook my head. It would still be a flour sack.

Oatmeal and cornmeal and salt came in cardboard cylinders just as they have for years. John and I watched as they grew empty, thinking of all the things we could make out of them. I thought about doll furniture or suitcases; John thought about drums and storage. John got the cornmeal boxes because cornmeal was used to make Johnny cake, his special favorite.

But how were things packaged in other stores where we shopped? We loved to watch the clerks wrap sweaters and "overhalls" and yard goods at J. C. Penney's or Montgomery Ward's or Damon's. They tore a sheet of brown paper off a huge roll, wrapped up the clothes and pulled a length of string down from a big ball above their heads. Then they expertly wound the string around the package both ways, broke it off with their bare hands and tied a double knot. We picked our package up by the string and proudly walked out of the store. We had plans for that paper and string, too. Kites.

If we bought shoes—one or two pairs a year for us kids because our feet were growing, a pair once in a great while for Mama or Pa. They said their feet

We were lucky to have a car during the Great Depression and World War II. It usually stayed in the garage during the week and only went to town on Saturday and to church on Sunday. Gas and tires were not to be wasted on "just driving around."

weren't growing anymore and they took good care of their shoes. The clerk just tied the string around the box. We were glad to hold it so that everyone could see we had new shoes. The box was a special bonus, too. We used it in our dresser drawers to keep socks and undies separated. We used it to store our collections. Or we used it to carry a couple kittens around—maybe in the bike basket.

Newspapers had lots of uses in our house. Every morning Mama started the fire in the cookstove with a wadded up newspaper, corncobs and kindling. (Corncobs and kindling are waste by-products of the farming industry now, but in our day, they were a major source of fuel.) After Mama fried beefsteak and poured out the gravy—Pa liked Mississippi gravy, which was just a little boiling water added to the frying pan after the meat was taken out—Mama wiped out the frying pan with a newspaper and tossed the paper in the cookstove.

When Mama washed the linoleum floor in the kitchen, she spread newspapers over the places we usually walked. They kept the floor clean a little longer. Later the newspapers went into the cookstove, too. Newspapers were great for cleaning windows. After we scrubbed the windows with a rag and ammonia (pew!), we wiped them dry with newspapers, and they sparkled in the sunshine. For any messy job we had—sorting through the nail can for just the right nails, cleaning out the "everything drawer," polishing shoes—the first step was spreading down a newspaper.

What about things people throw away when cooking in the kitchen, like potato and carrot peelings, apple cores and peelings? The chickens were glad of them. Mama bought a lug of peaches every fall. She dipped each peach in boiling water, slipped off the skin, cut it down the middle, took out the pit, and placed each half "just so" in the fruit jars so they looked like stacks of little caps. She finished the canning process in boiling water. Oh the jars looked so pretty lined up on the fruit cellar shelves. And the pigs slurped up the skins and crunched the peach pits—we liked to listen to them eat.

Table scraps went to the dog and cats. Not much was left on our plates after a meal, just bones or fat or gristle, maybe the tough meat that my little teeth couldn't chew, the toast crust that was too burned, the egg yolk that was too scary—I was afraid of food, not really a picky eater but very careful. I'd seen eggs with half-formed chicks and milk pails with flies floating on top.

The word "leftovers" was unknown in our house. Mama planned meals so that all food was eaten sooner or later. Roast beef left from dinner at noon turned up as hot-beef sandwiches or hash at supper. Boiled potatoes left from dinner became fried potatoes at supper. Fried chicken from dinner became chicken and homemade noodles at supper. Of course the noodles were homemade. Who ever would want Boughten noodles?!

Water was a precious commodity on our farm. The windmill by the barn had to be watched carefully to see that the cattle tank was full but not running over. That well provided our drinking water, too. A pailful had to be carried up to the house morning and night. We didn't want to waste it because that meant an extra trip to the well for one of us. Or we would just have to go thirsty.

Water for washing dishes or floors or clothes came from the cistern south of the house. We pumped it with a little hand-pump at the kitchen sink. This was rainwater that ran off the roof into the gutters, down the downspout and through an underground pipe into the cistern. We felt lucky to have a pump in the kitchen because we knew a lot of people had to carry cistern water into the house, too. Still, we were not wasteful of that water. Who knew if there would be enough rain in the summer to see us through the year? After Mama finished washing clothes, that water was used to wash the floor of the back porch and to wash the seat and floor of the backhouse and to wash the chicken manure off the sidewalk and to wash off our muddy boots.

None of that good soapy water went to waste on Monday morning, even after it had washed the line-up of a week's laundry from white sheets and Sunday shirts in the cleanest, hottest water and on to school clothes, work clothes and finishing with muddy overalls by Monday noon. Mama was tired out on Mondays.

What happened to our old worn-out clothes? They went into a basket by the sewing machine. The best parts of them showed up on the elbows or knees of our not-quite-so-worn-out clothes. We weren't embarrassed to wear patched clothes to country school. Everybody did. "Waste not, want not," we all said.

Other Books from South Bear Press

Our Natural Treasure: Genevieve Kroshus. Stories collected by Geraldine Schwarz. Life in Winneshiek County, Iowa, remembered with warmth and humor by a Norwegian American octogenarian. Designed by Jorunn Musil. 96 pp. 58 bxw photos. ISBN 978-0-9761381-4-3. Clothbound. $25 plus shipping.

Marguerite Wildenhain and the Bauhaus: An Eyewitness Anthology. Dean and Geraldine Schwarz, eds. Designed by Roy R. Behrens. As a potter, a teacher, a writer and a mentor, Marguerite embodied the best of Bauhaus standards in Germany and in her Pond Farm Pottery School in California. Her students, grandstudents and great-grandstudents carry on her traditions. 770 pp. 837 illus. (402 bxw, 436 color). ISBN 978-0-9761381-2-9. Clothbound. $75 plus shipping.

Adrift in Stormy Times. Rudolf Thill. Thill grew up in Germany, became a soldier, then a prisoner of war in the U.S. and France, and years later emigrated to Des Moines, Iowa, where he attended Drake University and taught at Grand View College. Designed by Zelda Productions, Decorah, Iowa. 262 pp. 32 bxw photos and illustrations. ISBN 0-9761381-0-7. Paperbound. $14 plus shipping.

MARGUERITE A Diary to Franz WILDENHAIN. Letters in the form of a diary by Marguerite Wildenhain. Dean Schwarz, ed. Designed by Deb Paulson with photographs by David Cavagnaro and woodblock prints by David Kamm. 125 pp. 70 illlus. (bxw and color). ISBN 0-9761381-1-5. Paperbound. $40 plus shipping.

Centering Bauhaus Clay: A Potter's Perspective. Dean and Geraldine Schwarz. Designed by Roy R. Behrens, 63 pp. 20 bxw illus. ISBN 978-0-9761381-5-0. Paperbound. $20 plus shipping.

South Bear Press books and Bobolink Books can be purchased at www.southbearpress.org or www.bobolinkbooks.com or go to Amazon.com (new and used).